Middle East Today

The Iranian Revolution of 1979, the Iran-Iraq War, the Gulf War, and the US invasion and occupation of Iraq have dramatically altered the geopolitical landscape of the contemporary Middle East. The Arab Spring uprisings have complicated this picture. This series puts forward a critical body of first-rate scholarship that reflects the current political and social realities of the region, focusing on original research about contentious politics and social movements; political institutions; the role played by non-governmental organizations such as Hamas, Hezbollah, and the Muslim Brotherhood; and the Israeli-Palestine conflict. Other themes of interest include Iran and Turkey as emerging pre-eminent powers in the region, the former an 'Islamic Republic' and the latter an emerging democracy currently governed by a party with Islamic roots; the Gulf monarchies, their petrol economies and regional ambitions; potential problems of nuclear proliferation in the region; and the challenges confronting the United States, Europe, and the United Nations in the greater Middle East. The focus of the series is on general topics such as social turmoil, war and revolution, international relations, occupation, radicalism, democracy, human rights, and Islam as a political force in the context of the modern Middle East.

More information about this series at
http://www.palgrave.com/gp/series/14803

Farhad Rezaei

Iran's Foreign Policy After the Nuclear Agreement

Politics of Normalizers and Traditionalists

palgrave
macmillan

Farhad Rezaei
Center for Iranian Studies in Ankara
Ankara, Turkey

Middle East Today
ISBN 978-3-030-08289-5 ISBN 978-3-319-76789-5 (eBook)
https://doi.org/10.1007/978-3-319-76789-5

Cover design by Thomas Howey

Printed on acid-free paper

This Palgrave Macmillan imprint is published by the registered company Springer International Publishing AG part of Springer Nature.
The registered company address is: Gewerbestrasse 11, 6330 Cham, Switzerland

This book is for you, Dad. As you look down from heaven, I hope you're proud of your son.

Contents

About the Author

Farhad Rezaei is a senior Iran analyst at the Centre for Iranian Studies (IRAM), Ankara, Turkey.

He is a member and contributor to the Association for the Study of the Middle East and Africa (ASMEA), and the Gulf 2000 Middle East Project at Columbia University's School of International and Public Affairs, New York.

Dr. Rezaei is the author of several books, numerous articles and published papers on a broad range of topics, including Iran's foreign policy, Iran's nuclear program, nuclear weapons proliferation and nuclear and radiological terrorism. These include 1) Iran's Nuclear Program: A Study in Proliferation and Rollback, New York: Palgrave Macmillan, 2017, and 2) Iran, Israel and the United States: The Politics of Counter-proliferation Intelligence, (co-authored), Lexington Books, Rowman & Littlefield 2018.

His articles have been published by Harvard–Iran Matters, the National Interest, Middle East Policy, *British Journal of Middle Eastern Studies*, *Journal of International Affairs*, *Insight Turkey*, *Korean Journal of Defense Analysis*, *Asian Affairs*, Atlantic Council, and The Hill, among others.

CHAPTER 1

The Negotiated Political Order and the Making of Iran's Foreign Policy

The political system created in the wake of the 1979 revolution has puzzled political experts and lay observers alike. In the first two decades of the Islamic Republic, analysts had struggled with finding a proper definition of the Iranian polity. A perusal of the substantial literature on the subject indicates that the definitions range from Islamofascism to democratic theocracy. Compared to the voluminous discourse on such definitions, there was little effort to understand the organizational structure of the new polity.

As a rule, state epistemology conceptualizes states as hierarchical structures, akin to a pyramid. In a hierarchical structure, a higher level indicates a greater measure of power and control over the lower levels. In a democracy, the executive and legislative branches occupy the apex; in a dictatorship, a single dictator or a small oligarchy wields absolute power. Either way, the chain of authority extends from top to bottom.

There have been only a few exceptions to this rule. In the dual state party system of the former Soviet Union, the hierarchical structure of the state nominally coexisted with the hierarchical structure of the party. In practice, however, the Communist Party hierarchy dominated its state counterpart. So much so that the Politburo, which elected the General Secretary, was the de facto pinnacle of the pyramid and the Secretary-General served as the head of state, the premier. This arrangement created occasional tension between the state's organs such as the Council of Ministers of the Soviet

F. Rezaei, *Iran's Foreign Policy After the Nuclear Agreement*, Middle East Today, https://doi.org/10.1007/978-3-319-76789-5_1

Union and the parastatal party bureaucracy. If anything, in Nazi Germany the state party duality was even more skewed toward the party that reigned supreme. In both cases, it was the party's chain of authority that dominated the entire political system.

There is little doubt that Ayatollah Ruhollah Khomeini, the founding father of the Islamic Republic and its first Supreme Leader, wanted to create a theocratic totalitarian state with clear authority lines. In his original 1941 treatise, Kashf al-Asrar (The Discovery of Secrets) and his 1975 opus Velayat-e Faqih, Khomeini advocated a numinous authority system—the divine rule by a religious guardian. A religious leader was necessary to turn the country into a "republic of virtue," where people, flawed by nature, could be perfected and achieve salvation. Khomeini insisted that only a strict religious theocracy could block the harmful secular messages and support Sharia-compatible behavior.[1]

But other leaders of the Islamic Revolutionary Party that guided the Khomenist movement understood that those who had rebelled against the Shah in the name of democracy would not embrace another dictatorship. Akbar Hashemi Rafsanjani, a close associate of Khomeini, urged to adopt a mixed system with divided legitimacy as defined by Max Weber. Part of the legitimacy is numinous, derived from the right of the God-anointed guardian, and part is legal-rational legitimacy that is reflective of the will of the people. In modern democracies, citizens express their will in periodic elections, but the electoral process in Iran is quite peculiar.

The Iranian president and the parliament, the Majlis, are ostensibly elected by a popular ballot. In practice, the Guardian Council, an appointed 12-member body of six faqihs and six jurors, exercises considerable control over the democratic process. The Council approves candidates for the presidency and the Majlis, and has the final say over its legislative output. Although considerably circumscribed, the elected officials preside over the statist part of the Islamic Republic, which includes the state bureaucracy, the budget, and the regular military, the Artesh, and other state institutions.[2]

But Khomeini and his aides understood that a statist structure could not protect their cherished revolution, a task that fell to the Islamic Revolutionary Guard Corps (IRGC), commonly known as the Revolutionary Guards. Established in May 1979, Article 150 of the Constitution described the

[1] Ruhollah Khomeini, Risaleh-yi Towzih al-Masa'il [Explanation of problems] (Qom: Ruh Press, 1947). (This edition also includes material written after 1947.)

[2] Hossein Bashirieh, A Preface to the Political Sociology of Iran of the Islamic Republic of Iran (Tehran: Negah Moaser, 2015).

Guards as "guardian of the Revolution and its achievements." Indeed, the IRGC took this mandate very seriously. In the words of one of its commanders, the Guards "possessed the moral essence of the revolution." The Basij (Sāzmān-e Basij-e Mostazafin), the Organization for the Mobilization of the Oppressed, was a voluntary militia in the Iran–Iraq war and subsequently incorporated into the Guards. After the end of the war, the Basij took a leading role in overseeing internal security, policing morals, and the suppression of dissidents.[3]

For all his moral authority, Ayatollah Khomeini's Trotskyite idea of a "permanent revolution," proved controversial. In a programmatic speech in February 1979, Khomeini announced his determination to export the revolution to neighboring, mostly Sunni countries and beyond. Radical "revolutionary exporters" associated with Ayatollah Ali Montazeri and his son Mohammad were eager to follow Khomeini's mandate. But their freelancing ventures in several neighboring countries alarmed the then Foreign Minister Ali Akbar Velayati, who urged a more disciplined approach. Ayatollah Rafsanjani, the budding pragmatist among the ideologues in Khomeini's entourage, was also concerned. After some organizational iteration, the consolidated Office of Liberation Movement (OLM) joined the Foreign Ministry.

From the very beginning, the IRGC demanded and received a large measure of autonomy from the state. Towards the end of the Iran–Iraq War, the regime's leaders realized the importance of reconstructing the country and maintaining domestic stability, notably by creating economic opportunities for Guards veterans. Rafsanjani facilitated the Guards' involvement in economic activities, and his privatization policy helped them to take control of several confiscated state companies. In 1989, the IRGC founded Khatam al-Anbiya Construction Headquarters (Gharargah Sazandegie Khātam al-Anbiyā), a company that subsequently branched out into industrial, agricultural, mining, road construction, transportation, and import and export sectors. When Ayatollah Ali Khamenei succeeded Khomeini as Supreme Leader in 1989, he insisted that the Revolutionary Guards should have priority in buying stock in the newly privatized companies, and he also exempted them from taxation.[4]

[3] Afshon Ostovar, Vanguard of the Imam: Religion, Politics, and Iran's Revolutionary Guards (London: Oxford University Press, 2016); Constitution of the Islamic Republic of Iran, rc.majlis.ir/fa/content/iran_constitution.

[4] Hashemi Rafsanjani, "Hashemi Rafsanjani: Our Deal was for Construction, not Ownership", *Hashemirafsanjani.ir*, December 16, 2014, http://hashemirafsanjani.ir/tax-

When Mahmoud Ahmadinejad came to power as president on a hardline ticket in 2005, his government dramatically increased support for the Revolutionary Guards. More than half of his cabinet members were either Guards veterans or people with ongoing ties to the organization. The Revolutionary Guards were awarded hundreds of no-bid government contracts and billions of dollars for construction and energy programs. To mask its extensive hold on the economy, the IRGC established hundreds of seeming private companies, which were run by a network of connected veterans. Unlike a traditional centrally administered organization, the IRGC economic enterprise was akin to a "pyramid ownership structure," where the original company acquired a new company by buying their shares or appointing members of the board, a process repeated numerous times as more companies joined the network.

Two larger foundations, Bonyad Mostazafan (Foundation of the Oppressed or the Mostazafan Foundation) and the Bonyad Shahid va Omur-e Janbazan (Foundation of Martyrs and Veterans Affairs), and several smaller ones cemented the parastatal domain. The foundations commanded enormous wealth, most of it from former the Pahlavi Foundation and the confiscated property of wealthy Iranians who fled after the revolution. In addition to their charitable functions, the foundations have operated a network of businesses, some with connections to the Revolutionary Guards. For instance, the Bonyad Shahid Foundation has a strong link to the Revolutionary Guards; it has recently appointed the former Guard's Air Force commander Hossein Dehghan as president.[5] Bonyad Shahid has large industrial ventures, including the Industrial Development and Renovation Organization of Iran, known as IDRO Group, was established in 1967 during the Shah's rule.

While the state structure had evolved a semblance of hierarchy, the parastatal domain has continued to be quite chaotic, especially in the initial post-revolution period. In battling for ideological influence, power, and resources, the parastatals not only impeded good governance but made governing difficult altogether. Rafsanjani, who emerged as a key player in the new regime, had soon realized the danger of out of control parastatals. He tried to pressure the Revolutionary Guards to merge with the Artesh, a move that the commanders rejected out of

onomy/term/267; Farhad Rezaei & Somayeh Khodaei Moshirabad, The Revolutionary Guards: From Spoilers to Accepters of the Nuclear Deal, *British Journal of Middle Eastern Studies*, Vol. 43. No. 4. 2016.

[5] FNA, Bonyad Mostazafan: From Establishment to Now, *Fars News Agency*, September 8, 2014, www.farsnews.com/newstext.php?nn=13930617001295.

hand. Rafsanjani was more successful with the Basij, which was ordered to join the Revolutionary Guards in following a brutal purge; the OLM was forced out from the Foreign Ministry and merged with the Quds Force (QF), the foreign operation branch of the IRGC. Still, the intense rivalry within and among the parastatals took decades to subside. A 2009 Rand study commissioned by the U.S. Department of Defense found that factionalism was quite prevalent. As the authors of the study put it, "These early instances of sporadic and uncontrolled factionalism established a pattern within the IRGC that continues to this day."[6]

More problematic from the governance perspective was the failure to clarify the relations between the state and the parastatals. To solve the problem, Ayatollah Khomeini had declared himself as the Supreme Leader, a position that gave him the power to solve the frequent conflicts between the state and the parastatals. But in reality, this organizational innovation was less than optimal because of Khomeini's failings. A growing body of work indicates that the Supreme Leader, in poor health and seemingly isolated in his quarters, was inconsistent in his attitudes and opinions at best and confused at worse. Manipulating him or going behind his back became the accepted modus operandi in Tehran. When the activists from the Students Following the Line of Imam broke into the American embassy on November 7, 1979, and detained the diplomats, Khomeini was initially reluctant to bless the operation. However, after the militants implied that he lacked revolutionary fervor, the Ayatollah changed his position two days later despite opposition from Rafsanjani.[7]

Facing a bloody war with Iraq, Rafsanjani and the then president Ali Khamenei, concluded that Iran needed nuclear weapons. Rather than fight Khomeini, who issued a fatwa declaring nuclear weapons to be un-Islamic, they and other leaders of the so-called "nuclear sanctum" decided to start the project without notifying the Supreme Leader.[8]

While Ayatollah Khomeini was alive, little could be done to streamline the rather chaotic performance of the Supreme Leader in his role as an arbiter of conflicts between the state and parastatal intuitions of the regime; it was his successor, Ali Khamenei, who institutionalized the role, laying the foundation for the negotiated political order.

[6] Frederic Wehrey et al., The Rise of the Pasdaran: Assessing the Domestic Roles of Iran's Islamic Revolutionary Guards Corps, RAND, 2009, 81.

[7] Ofira Seliktar, Failing the Crystal Ball Test: The Carter Administration and the Fundamentalist Revolution in Iran (USA: Praeger, 2000).

[8] Farhad Rezaei, Iran's Nuclear Program: A Study in Proliferation and Rollback (New York: Palgrave Macmillan, 2016).

The Negotiated Political Order: A Precarious Balancing Act

After Khomeini's choice of successor, the highly regarded Grand Ayatollah Hussein-Ali Montazeri, broke ranks with the regime, the inner circle settled on Ali Khamenei. The new Supreme Leader lacked the necessary scholarly qualifications but was an adept politician capable of maneuvering the complex relations between the state and the parastatals. In a move befitting a seasoned bureaucrat, Khamenei set up the Rahbar's Office staffed with aides who liaised with the various branches of the government, the Revolutionary Guards, and the foundations. Currently, the Supreme Leader holds the chair of the Supreme National Security Council (SNSC), the commander of the armed forces, and personally oversees the QF, an arrangement that gives him a huge say in matters of revolutionary export. Perhaps most importantly, the Supreme Leader is said to control the Execution of Imam Khomeini's Order (Setade Ejraiye Farmane Emam) or Setad, a parastatal foundation with an estimated worth of $95 billion in 2015. Because of the foundation, Khamenei is free from the constraints of the government and the budget, a position that has made him an independent player in a range of legitimate and clandestine activities. Scholars who have studied Khamenei have been impressed with his alleged control of the political system. As one prominent expert put it, the Supreme Leader is an "omnipotent overseer of Iran's political scene."[9]

Despite such lofty descriptions, the extent of Khamenei's ability to act as the ultimate arbiter of the political systems is not entirely clear. Far from being the "omnipotent overseer," the Supreme Leader has had to navigate the complex labyrinth of competing factions and engage in ad hoc coalitions with those who could guarantee his political survival and that of the regime. Unlike the highly respected Montazeri who commanded a wide following, the modestly credentialed new leader needed a support base to prevail. Having been involved in the Iran–Iraq war and the nuclear project, Khamenei has tilted toward the Revolutionary Guards. In fact, the Guards saved him for serving in a triumvirate, an idea which Rafsanjani and others supported to get around the issue of credentials.

Khamenei's reliance on the Guards and the Basij deepened after the moderate reformer Seyyed Mohammed Khatami won in a landslide in the 1997 presidential election. Having defeated the hardline Ali Akbar Nateq Nuri from the Combatant Clergy Association (Jāme'e-ye Rowhāniyyat-e

[9] Vali Nasr, "Meet the Decider of Tehran," *Washington Post*, December 9, 2007.

Mobārez) and Coalition of Islamic Societies (Ḥezb-e moʾtalefa-ye eslāmi), the preferred candidate of the Revolutionary Guards, the coalition that brought Khatami to victory pushed for a "second revolution," a code for reforming the political system. Key among the plethora of changes that the reformers demanded was the curtailing of the parastatals and their economic empire. The winning Khatami coalition preferred a clear chain of state authority to what they considered to be a convoluted system chaired by the Supreme Leader-cum-referee. But to the Supreme Leader and his patron, the Revolutionary Guards, Khatami's demands presented an existential threat.

Having determined that overthrowing Khatami would be politically dangerous, Khamenei and the Guards settled on a sophisticated formula of hollowing out the power of the state institutions. In swift order, the Rahbar's Office and parastatals usurped or duplicated many of the state functions. In April 1998, Major General Yahya Rahim Safavi, the IRGC commander, declared his intention to "break up the pens and cut out the tongues" of reformist intellectuals and journalists. As if on cue, the Basij and several shadowy auxiliaries launched coordinated violent attacks on meetings of Khatami's supporters. In the same year, a few prominent intellectuals were killed, and others went missing. Unable to protect his followers, the president had repeatedly appealed to the Supreme Leader who, by some accounts, was supporting the parastatals. Between 1998 and 2001, the Guards, working in tandem with the Ministry of Intelligence and Security (MOIS) (also known as Vezarat-e Ettela'at va Amniyat-e Keshvar or VAVAK), ostensibly under the control of the Khatami government, undermined the reform movement.

One incident was highly indicative of the ability of the parastatals to confront the government. On February 11, 2004, the Turkish company Tepe-Akfen-Vie won a Ministry of Communication bid to operate the new Imam Khomeini Airport. But on May 18, 2004, on the first day of operations, the Guards air force abruptly shut down the new venue and diverted all the incoming flights to Isfahan International Airport and Mehrabad Airport respectively. The incident created an international scandal and strained Iranian–Turkish relations, but the Supreme Leader did not intervene. On the contrary, the Guards, whose bid for the same contract had been rejected, won a resounding victory when the Minister of Transportation was impeached.[10]

[10] Frederic Wehrey et al., The Rise of the Pasdaran: Assessing the Domestic Roles of Iran's Islamic Revolutionary Guards Corps, RAND, 2009, 74; TAV: We trust Iran enough to carry

The backlash against Khatami's plan to reform the political system spurred other parastatal groups into action. Among the most important was the Haqqani School, an ultra-conservative network under the leadership of Ayatollah Mohammed Taqi Mesbah Yazdi. Ayatollah Ahmad Jannati was another prominent leader in the network, whose alumni had an extensive presence in the clerical, military, and judicial system. Hardline Haqqani-affiliated jurists including Jannati, were behind the novel doctrine that it is the role of the judiciary rather than the state to target individuals who embraced liberalism. To that end, Ayatollah Mahmoud Hashemi Shahroudi, the Chief Justice who espoused the Haqqani circle ideas, had prosecuted several liberal Majlis members despite their immunity. President Khatami complained, but the Supreme Leader, who had appointed Shahroudi, sided with the judiciary.

Because of the failure to develop a clear authority structure, the various players in the system had to resort to a peculiar form of a negotiated order. Whereas classic style negotiations are defined as a process where two or more parties with competing interests come together to reach an agreement, the regime actors tend to power posture to affect the subsequent outcome. Though normally less dramatic than the airport closure, such tactics are quite normal. For instance, the judiciary, acting in collusion with MOIS or the Basij, had arrested individuals, often on bogus charges, to pressure the government. Khatami, who famously promised a Dialogue Among Civilizations and sought a detente with the United States, was particularly vulnerable to power plays of his opponents.

A system that featured multiple players with no clear rules of engagement and no firm authority system was at the mercy of spoilers, defined as "leaders and parties who believe the emerging peace threatens their power, worldview, and interests and who use violence to undermine attempts to achieve it." Acting as spoilers, even relatively weak actors could undermine an arrangement agreed by the state and parastatals. A disgruntled faction or factions could torpedo an otherwise agreed-upon plan of action, as can be seen in the case of the Shirazi Jews. In April 1999, MOIS arrested 23 Jews from Shiraz, Fars province and put 13 of them on trial. Ayatollah Jannati threatened that, if found guilty, they would be executed; a trans-

out business there, February 10, 2004, www.hurriyetdailynews.com/tav-we-trust-iran-enough-to-carry-out-business-there.aspx?pageID=438&n=tav-we-trust-iran-enough-to-carry-out-business-there-2004-02-10; *Andrew Higgins*, A Feared Force Roils Business in Iran,October17,2006,www.ncr-iran.org/en/news/terrorism-fundamentalism/2382-a-feared-force-roils-business-in-iran.

parent effort to sabotage Khatami's detente with the United States, where the case received considerable publicity. After a lengthy negation with the judiciary, the Jews were spared their lives, but some languished in jail for more than a decade.[11]

Because of the constant coalitional realignments between the various factions, the negotiated political order is always in a certain amount of flux. Even astute observers were frequently unable to detect changes and at a loss to explain some sudden developments. Nowhere was the disconnect between the surface appearance and underlying realities more apparent than during the second term of President Khatami. It was during this period that a virtually ignored change had taken place in the realm of state and parastatal relations.

Jolted by Khatami and the reformist movement, in the early 2000s, a hardline coalition named the Builders of Islamic Iran (Etelaf-e Abadgaran-e Iran-e Islami), or Abadgaran had emerged. Boasting Ayatollah Mohammed Taqi Mesbah Yazdi as its spiritual leader, Abadgaran attracted hard-core personalities. Ayatollah Abul-Ghasem Khazali, the head of the secretive messianic Mahdavit society, the Awaited One Foundation (Al Ghadeer Foundation), the commander of the Ansar al- Hezbollah, Brigadier General Mohammed Hejazi, the Basij chief, and Mojtaba Khamenei, the son of the Supreme Leader were among them. Equally important, Rahim Safavi and his deputy, Mohammed Baqer Zolgader, were fully supportive of the new political party.

Mahmud Ahmadinejad, a little-known politician and a former Guards member, joined the group in 2002 as part of the Islamic Society of Engineers. An essentially revivalist movement, known as Principalism, the Abadgaran wanted to restore the original revolutionary vision of Khomeini, remove Western values of liberalism, democracy, and end the corruption of the clerical classes. After Abadgaran had won a string of local elections in 2003, the Abadgaran- dominated city council in Tehran nominated Ahmadinejad as mayor in 2004. Using his new perch, Ahmadinejad cultivated the Guards, mainly by awarding their companies no-bid tenders for municipal projects. When Ahmadinejad won the presidential election in 2005 even Iranian observers were stunned. The Western media and the pundits scrambled to describe the new president and his conservative Principalist beliefs.

[11] Stephen John Stedman, "Spoiler Problems in Peace Processes," *International Security* 22 (Fall 1997): 5–53.

The new president wasted little time in elevating the relations with the IRGC to a new level. He appointed many former Guards officials to his cabinet and offered lucrative no-bid tenders to Guards-controlled companies. The government awarded a $1.2 billion contract to Khatam al-Anbiya to develop the giant Pars oil field and a $2 billion contract to build the Tehran metro. In fact, between 2005 and 2009, the estimated worth of the Guards ventures went from some $50 billion to $90 billion.

The close relations between the president and the IRGC upended the negotiated political order. For once, a coalition of the state and the parastatals forces found themselves fighting the reformists whose position was marginalized, not least because MOIS, the Guards security forces, and the independent judiciary launched a wave of persecution against dissidents. In a sign of the newfound state-parastatal comity, the government discontinued all efforts to police the Guards' smuggling operations, an activity that, according to the Khatami administration, cost the state millions of dollars in lost duties. The Supreme Leader, severely shaken by the reformists' efforts to undermine the theocracy, was pleased by the nexus of the Guards, Basij, and Ansar-e Hezbollah, which imposed an Islamic-appropriate lifestyle on the population, after years of relaxation under Khatami.

Unsurprisingly, the Guards decided to support Ahmadinejad in the 2009 election even though their opinion polls indicated a resounding victory for the Green Movement reformers. A few hours after the polls closed, the Supreme Leader declared Ahmadinejad as the winner. In the wake of the widespread riots and their brutal suppression by the Guards auxiliaries and security apparatus, many observers predicted the end of the negotiated political order. They suggested that Iran was turning into a military dictatorship with a thin theocratic veneer. Noting that Tehran "is increasingly relying on the military to control its citizens," the respectable *Foreign Policy* journal predicted the rise of an Iranian dictatorship.[12]

But Ahmadinejad, the beneficiary of the IRGC's intervention, turned out to be increasingly volatile and erratic, a behavior which unnerved the Supreme Leader. Although a founding member of the "nuclear sanctum" and a staunch supporter of the nuclear program, Ayatollah Khamenei was concerned by Ahmadinejad's nuclear defiance which, by 2010, prompted the United Nations to impose increasingly punishing sanctions on Iran. With the economy in a free- fall, some in the Rahbar's Office network came to question the president's foreign policy and even his common sense. Ali Larijani, a protégé of Khamenei and the secretary of the SNSC,

[12] Geneive Abdo, "The Rise of the Iranian Dictatorship," *Foreign Policy*, October 7, 2009.

and who also served as the top nuclear negotiator, spoke for the Supreme Leader when publicly criticizing Ahmadinejad's foreign policy style. The president promptly fired him, opening a rift with Khamenei and his foreign policy adviser, the former Foreign Minister Velayati. In December 2010, Ahmadinejad fired another ally of Khamenei, the Foreign Minister Manouchehr Mottaki who was abroad at the time. Things came to a head when, in April 2011, Ahmadinejad forced out the powerful head of MOIS Heydar Moslehi, another protégé of the Supreme Leader. Khamenei overrode the decision and reinstated Moslehi, a move which prompted Ahmadinejad to boycott cabinet meetings. This sensational development generated numerous but hard to substantiate rumors. One intriguing claim held that the president wanted to take over MOIS to search for evidence incrementing the Supreme Leader in alleged corrupt practices in the Mustazafin and Setad foundations. Ahmadinejad himself implied that Khamenei was usurping the decision-making of the government and targeting his trusted adviser and chief of staff Esfandiar Rahim Mashaei.[13]

Whatever the truth, the battle with the Supreme Leader could not have come at a worse time for Ahmadinejad. Trying to fight the deteriorating economy, the president revived Khatami's old campaign to bring the Guards' economic empire under state control. Because of the sanctions, the IRGC extended its smuggling operations which robbed the state of much-needed revenue. For instance, one report indicated that the state's customs and tax revenues decreased by $8 billion due to smuggling.[14] In a seemingly imprudent move, the president took the campaign public, referring to the Guards as "my smuggler brothers." Despite this open defiance, the Rahbar's network and the Guards decided that deposing Ahmadinejad would hurt the legitimacy of the regime. To prevent a similar situation, they also tried to find a more pliant candidate to compete the presidential election scheduled for 2013.[15]

[13] Aftab News, Ahmadinejad Accessed to First Class Documents in Ministry of Intelligence, June 14, 2016, aftabnews.ir/fa/news/374764/دست‌یابی-احمدی‌نژاد-در-وزارت‌اطلاعات-به-پرونده%E2%80%8Eهایی-درجه‌یکتخلف-خانه‌نشینی-احمدی‌نژاد-محرز-شد-اما-پیگیری-پرونده-را-به‌مصلحت-ندانستند; Babak Dad's Letter To Ministry of Intelligence About Stealing Classified Documents by Ahmadinejad, news.gooya.com/2017/03/post-2162.php; Ahmadinejad's Surprising Disagreement with Supreme Leader, June 27, 2015, www.mashreghnews.ir/news/433739/مخالفت-شگفت-انگیز-احمدی-نژاد-با-دستور-رهبری.

[14] ISNA, Smuggling and its Eight Billion Dollars Damage to the National Economy, *Iranian Students News Agency*, www.isna.ir/news/95040107613/درنگی-در-قاچاق-کالا-و-خسارت-هشت-میلیارد-دلاری-آن-به-اقتصاد.

[15] Entekhab, Does Ahmadinejad Began his Revelations? www.entekhab.ir/fa/news/30314/منظور-احمدی-نژاد-از-برادران-قاچاقچی-و-اتصال-شبکه-هاي-قاچاق-به-صاحبان-قدرت-و-نفوذ-چه-بود.

Unfortunately for them, the sanctions-induced economic breakdown turned the election into the referendum on the nuclear program. Reflecting a new political reality, a coalition of former Khatami reformers and Rafsanjani pragmatists under Hassan Rouhani won handily over Saeed Jalili, a traditionalist Principalist, the preferred candidate of the Supreme Leaders and the Guards. Rouhani's tenure marked the return to the Khatami-era negotiated political order with the state struggling with the parastatals over a broad range of issues. But this time around, the Supreme Leader, scarred by the legitimacy crisis which the sanction triggered, reluctantly supported Rouhani.

With the backing of the Supreme Leader Rouhani could negotiate the nuclear agreement formally known as the Joint Comprehensive Plan of Action (JCPOA) in July 2015, which offered sanction relief in exchange for giving up proliferation. Improved economics propelled Rouhani to a second victory in 2017, giving the moderate statists more political legitimacy. Indeed, the president was quick to seize on his popular mandate to claim that it was "the people's will" which should matter, not just the divine Velayat-e Faqih principle. Even more astounding, the government decided to take on the historically entrenched economic empire of the IRGC. According to reports, Rouhani persuaded Ayatollah Khamenei that the Iranian economy had reached a standstill because of the vast corruption, mismanagement, and intimidation in the companies run by the Guards. As Rouhani put it, the Guards created a "government with a gun" that "scared" the private sector. Although discreet, the government effort to curtail the parastatal economy has been unmistaken, with several high-ranking Guards arrested for embezzlement and corruption. While is not entirely clear whether Rouhani will succeed, he "is standing firm to bring the Guards under the general umbrella of the economy and give them projects only under certain competitive conditions," according to one observer.[16]

Even this brief analysis of the negotiated political order indicates the complexity of foreign policy decision-making in Iran.

Foreign Policy Decision-Making: Between Idealism and Realism

As part of the challenge to good governance, the negotiated political order increased the complexity of foreign policy making at the paradigmatic level. As is well known, realism and neorealism have competed with strands

[16] Asharq al Awsat, "Iranian Revolutionary Guard Corruption Empire," *Asharq al Awsat*, September 16, 2017.

of idealism in shaping states' view of the international domain. Realism and neorealism view the world as inherently dangerous, either because of human nature or the anarchic state of relations among countries. States that strive to survive in this environment and protect their interests need to be realistic and pragmatic.

Idealism, on the other hand, holds that the state's internal political philosophy should guide its foreign policy goals. Ironically, idealism has been associated with the so-called Wilsonian idealism, which stipulates that states have good intentions and working together could lead to a better, that is, more democratic, and peaceful world. Wilsonian idealism suffered a blow after the demise of the League of Nations, but its elements surfaced in the liberal strands of international theory that became popular among the architects of the post-WWII international world.

While realism-neorealism and idealism offer a sharply different concept of international reality, they do share the assumption that states are unitary actors. In the case of realism- neorealism, the state is also considered to be a rational actor. In his classic *Essence of Decision*, Graham Allison conceptualized the state as a monolithic decision-making unit, capable of setting goals and ranking them, considering options, assessing consequences, and engaging in profit maximization. Understandably, idealism does not offer such rigid stipulations, but it implies that a state is a unitary actor.[17]

But as noted above, the negotiated political order is difficult to square with the unitary actor concept embedded in International Relations (IR) theories. One way to address the problem is to identify elements of idealism and rationalism in the foreign policy thinking of the three sectors that make up Iran. With the exception of the Ahmadinejad period, idealist thinking has been more prevalent in the parastatal domain, whereas realism resided in the domain of the state. Ayatollah Khomeini tended to idealism, but under his successor, the Rahbar's office zigzagged between idealism and realism, reflecting Khamenei's strong instinct to assure the survival of the regime.

Khomeini's version of foreign policy idealism was linked to his view that the Iranian state was mandated to create the "republic of virtue" and, equally important, promote the idea abroad. Exporting the revolution was part of a broader critique of the of the West and its international institutions. Jalal Al-e-Ahmad, a popular intellectual, rejected the Western culture

[17] Graham Allison and Philip Zelikow, Essence of Decision: Explaining the Cuban Missile Crisis (Pearson, 1999).

and accused Iranians of "Westoxication," a form of corrupt Western identity. Ali Shariati, the highly influential political sociologist, and thinker, suggested basing the international order on a synthesis of Islamic and socialist values, a theme that the regime would later use in its so-called Third Way economy. Ahmadinejad, whose affinity to Shariati was well known, criticized the international system for its "colonial" and "imperialist" treatment of Muslim countries in general, and particularly Iran.[18]

By attacking the Western-oriented international order, Khomeini and his disciples hoped to justify the violation of international norms which the revolutionary export required. Indeed, the Revolutionary Guards and Quds Force fashioned their doctrine on the work by Brigadier General S.K. Malik, known as the "Islamist Clausewitz" of the Pakistani high command. Malik brought coherence to these disparate passages and rulings in his paradigmatic book *The Quranic Concept of War*, published in Lahore in 1979. Malik held that the Quranic imperative of jihad—a duty to wage war on non-Muslims—was to be shared by the entire community, not just by formally designated combatants. Malik saw no distinction between enemy combatants and non-combatants either.[19]

The IRGC-QF had institutionalized Khomeini's idealism and the Quranic concept of war in its doctrine. According to Hassan Abbassi—nicknamed "the Iranian Kissinger"—the head of the aptly named Center for Borderless Security Doctrinal Analysis at the Guards' Imam Hossein University, borders and national sovereignty was a Western invention alien to the spirit of the Muslim Ummah (nation). In his view, the colonial powers that divided the Middle East created grave incongruities between ethno-sectarian, tribal, and national entities. It was the duty of the Iranian Islamists to challenge these "tribes with flags" and fashion a new overall Muslim identity built on the Velayat-e Faqih concept. Utilizing the religious and ethno-tribal incongruities, the IRGC-QF became part of the revolutionary export strategy. Planting cells among sympathetic Shiite populations was a favorite tactic. Another tactic aimed at destabilizing the political system of target countries, either through terror attacks or civil unrest or a combination of both. Over time the QF, which shouldered the bulk of this mission, created eight directorates: Western countries, Iraq,

[18] Jalal Al-e-Ahmad, Gharbzadegi (Tehran: Mazdâ Publishers 1982); Ali Shariati, Martyrdom: The Islamic Renaissance Series (Tehran: The New Muslim View, 1973); Mahmoud Ahmadinejad, in meeting with IRGC Commanders, August 28, 2007.

[19] S.K. Malik, The Quranic Concept of War (Lahore, Reprint in India: Adam Publishers and Dirtributors, 1979).

Israel–Lebanon–Jordan, Afghanistan–Pakistan–India, Turkey, North Africa, Arabian Peninsula, and the former Soviet republics. Several Shiite proxies such as Hezbollah in Lebanon and the Houthis in Yemen have acted as force multipliers for the IRGC-QF and its "coercive influence" network.[20]

Khomeini's idealism did come with a price. In 1984 the United States designated Iran a state sponsor of terrorism and sanctioned the Revolutionary Guards and the QF as well as individuals and companies affiliated with them as a "foreign terrorist organization" (FTO). Iran's terrorism, which included the killing of Iranian dissidents in Europe in the 1980s and 1990s, contributed to the perception of the regime as a rogue. The European Union sanctioned Iran as a terrorist state, and the United Nations Security Council (UNSC) joined in the condemnation.

While declarations of fealty to Khomeini's idealism were de rigueur among the ruling class, the high opportunity cost of revolutionary export rankled many. No one was more aware of the problems than Ayatollah Rafsanjani, one of the most powerful politicians in the history of the Islamic Republic. After a series of executive and legislative positions, Rafsanjani served as president between 1989 and 1997. Unlike many of his revolutionary cohorts, Rafsanjani realized that the regime needed to adopt a more pragmatic international policy to survive and prosper. In the words of one analyst, the pragmatists had little use for a "phantasmagorical struggle with the immoralist Satan." But, as noted above, Rafsanjani failed to curb the Guards who objected to his modes of market reforms and his putative outreach to the West.[21]

His successor Khatami was only marginally more successful in launching a robust realist foreign policy. Despite the reformists' electoral success, the parastatals and the Rahbar's network, augmented by the Haqqani School activists, reacted quite violently to the Dialogue Among Civilizations. Rafsanjani, still politically active, concluded that the Iranian realists needed to build a grass-roots ideological platform. Hassan Rouhani, one of the early revolutionary leaders and the chair of the Foreign Relations

[20] Rajanews, Gozareshi az Taasise Sepahe Pasdaran-e Enghelab-e Eslami (A Report of Establishing of Islamic Republic Revolutionary Guards), *Raja News*, April 20, 2011, http://www.rajanews.com/news/69143; Afshon Ostovar, Vanguard of the Imam: Religion, Politics, and Iran's Revolutionary Guards (London: Oxford University Press, 2016).

[21] Islamic Republic of Iran Constitution, rc.majlis.ir/fa/content/iran_constitution; Afshon Ostovar, Vanguard of the Imam: Religion, Politics, and Iran's Revolutionary Guards (London: Oxford University Press, 2016).

Committee in the Majlis, formed a parliamentary group called Tadbir (Prudent) in 1998. The Prudent parliamentarian pushed for a realistic-pragmatic foreign policy, along the lines that Rafsanjani had advocated. Two years later, Rouhani formed the Party of Development and Moderation (Hezb Etedal va Tose), which called for the normalization of Iran's international standing. Mir Hossein Mousavi, who ran on the Party's platform in 2009, turning it into the Green Movement, demonstrated how popular the Rafsanjani–Rouhani brand of normalization was. To the parastatals and the Supreme Leader, however, realism was an anathema that produced the "poisoned fruit" of normalization.

Being robbed of electoral victory, the moderates had to wait for an opportune moment to push for normalization. It was Ahmadinejad's high-profile foreign policy idealism that created the breakthrough. Rhetorically, the president signaled a return to Khomeini's battle with the imperialist West as epitomized by the United States and Israel. His nuclear defiance brought sanctions which, as indicated, wrecked the Iranian economy. Unbowed, Ahmadinejad attacked the Non-Proliferation Treaty (NPT) as a colonial apartheid system that separated the nuclear "haves" from the "nuclear have-nots." He also described the International Atomic Energy Agency (IAEA) as a handmaiden of Western nuclear interests. With the economy in tatters, the 2013 election was a referendum on the cost of idealism in Iranian foreign policy.

Rouhani's victory and the subsequent JCPOA of July 2015 was a triumph for the normalizers. But given the Principalist objections, the implementation of the normalization project has not been easy.

Beyond the JCPOA: Normalization and Its Challenges

Improving the economy and integrating Iran into the community of nations has been a major challenge. When running for office in 2013, President Rouhani promised to reform the economy, ravaged by years of mismanagement, corruption, populism, and sanctions. Indeed, the hope of economic improvement drove the popular support for the nuclear deal. Within days of signing the JCPOA, President Rouhani declared in an open letter to First Vice President Eshaq Jahangiri his intention to heal the economy by focusing on the banking system. The pledges were popular enough to give Rouhani an edge in the 2016 Majlis election.

After analyzing the problems, the Rouhani government concluded that the cornerstone of economic recovery should be reform of the banking system, which had operated for decades "with low capital-adequacy requirements and inadequate regulatory and supervisory mechanisms." As a result, the banks were hobbled with a large inventory of nonperforming loans and lacked capital for extending credit to businesses. Valiollah Seif, the governor of Iran's Central Bank, commented that financial mismanagement had left Iran's financial system in "tatters." The Banking Overhaul Plan, signed into law by Rouhani in July 2016, was designed to remedy the situation. An official government website promised "to get financing for short- and medium-term projects back on track, provide a cash cushion to tackle bad loans, promote competition, reorder the money market by regulating the army of uncertified credit and financial institutions, and increase banks' lending power by raising their capital."[22]

Much as the law was appreciated, it did not go far enough to secure Foreign Direct Investment (FDI). Indeed, the custodians hoped to cash in on the JCPOA's promise of FDI, especially as the country's critical oil and gas sectors had become badly degraded under years of mismanagement and sanctions. But, as Seif put it, the "Iranian banks are outdated" for attracting FDI; they are "behind international norms and standards." The Central Bank governor's somewhat oblique reference pertained to the 1999 UN International Convention on the Suppression of the Financing of Terrorism.[23]

Despite the public posturing, behind-the-scenes maneuvering revealed a more nuanced reality. Far from assuming the role of an unmitigated spoiler, the Revolutionary Guards have been forced to engage with the Rouhani administration over a whole range of issues. On terror financing, the government presented the Revolutionary Guards and the Supreme Leader with a fait accompli. On June 12, 2016, Iran joined the Eurasian Group, a Financial Action Task Force (FATF)- style regional body that is expected to assist with the implementation of anti-laundering laws. The British authorities have also stepped up their help to Rouhani by providing reform assistance to three banks, Persia International Bank, Bank Melli Iran (BMI), and Sepah Bank. The BMI and Sepah Bank, once the premier outlets for illicit transactions, announced that they would not service Revolutionary Guards accounts, including that of the giant construction concern

[22] "Rouhani Signs Banking Loan into Law," *Financial Tribune*, July 2, 2016, financialtribune.com/articles/economy-business-and-markets/44712/rouhani-signs-banking-reform-plan-law.

[23] Ibid.

al-Anbiya.[24] Some Tehran insiders predicted that the Rouhani administration would force other banks to comply with FATF-style policies, limiting the Revolutionary Guards' access to banking services.[25]

While an impediment to the normalization project, the banking issue was minor compared to the problem that the Rouhani moderates have had to face in navigating a variety of issues pertaining to normalization, including terrorism and human rights. As the following chapters illustrate, Iran's foreign policy serves as a battleground between the moderate normalizers and the traditional Principalists.

Conclusion

Unlike the standard hierarchical polity or even the dual-party state system in the former Soviet Union, Iran's Negotiated Political Order is based on a series of complex arrangements among elites whose power base is anchored either in the state or parastatal domain. The president and the state bureaucracy compete with the parastatal Revolutionary Guards, the large revolutionary foundations and the ultra-conservative clerics.

In principle, the Supreme Leader is tasked with making binding decisions, but his power is far from absolute. Secretive, complex, and intense maneuvering, intimidation, brinkmanship, and even violence combine to generate a fluid, and opaque decision-making process. This often results in contradictory messages emanating from Tehran.

Both domestic and foreign policy issues have fallen prey to the fierce and sometimes violent competition among the elites. Because the Guards, the bonyads, and the clerics operate outside the state, they feel little need to follow the decisions of the state institutions, including the government. The Guards are notorious for espousing foreign policy goals that clash with the position of the president and his administration. Occasionally, the parastatals even sabotage the decisions made by the government, a pattern most prevalent on issues pertaining to terror and human rights.

[24] FNA, "Iran Joined the Eurasian Group," *Fars News Agency*, June 12, 2016, www.farsnews.com/13950323000767; Martin Arnold, "British Regulators Help Iranian Banks Come in From the Cold," *Financial Times*, January 31, 2016, http://www.ft.com/cms/s/0/54144ec2-c692-11e5-b3b1-7b2481276e45.html.

[25] ANA, "Government Under Attack for Implementing FATF," *Aftab News Agency*, September 6, 2016, http://aftabnews.ir/fa/news/394086/; 9Sobh, What the FATF Solicitous (Delvapasan) Are Looking for, *9Sobh News Agency*, September 23, 2016, 9sobh.ir/fa/news/95839/به-دنبال-چه-هستند-fatf-دلواپسان.

Although the differences between the normalizers and the Principalists are ideological, the sectoral overlap is very significant. The normalizers represent the state sector and the modest civil society, which fights the Islamist theocracy. The Principalists reflect the interests of the Revolutionary Guards, the Basji, and the disciples of the Haqqani School who want to eliminate all republican-democratic elements in the Constitution.

References

9Sobh, What the FATF Solicitous (Delvapasan) Are Looking for, *9Sobh News Agency*, September 23, 2016, 9sobh.ir/fa/news/95839/به-دنبال-چه-هستند-fatf-دلواپسان.

Afshon Ostovar, Vanguard of the Imam: Religion, Politics, and Iran's Revolutionary Guards (London: Oxford University Press, 2016).

Aftab News, Ahmadinejad Accessed to First Class Documents in Ministry of Intelligence, June 14, 2016, aftabnews.ir/fa/news/374764/دستیابی-احمدی‌نژاد-در-وزارت‌اطلاعات-به-هایی-درجه‌یک‌تخلف-خانه‌نشینی-احمدی‌نژاد-محرز-شد-اما-پیگیری-پرونده‌را-به‌مصلحت-ندانستند%E2%80%8Eپرونده.

Ahmadinejad's Surprising Disagreement with Supreme Leader, June 27, 2015, www.mashreghnews.ir/news/433739/مخالفت-شگفت-انگیز-احمدی-نژاد-با-دستور-رهبری.

Ali Shariati, Martyrdom: The Islamic Renaissance Series (Tehran: The New Muslim View, 1973).

ANA, "Government Under Attack for Implementing FATF," *Aftab News Agency*, September 6, 2016, http://aftabnews.ir/fa/news/394086/.

Andrew Higgins, A Feared Force Roils Business in Iran, October 17, 2006, www.ncr-iran.org/en/news/terrorism-fundamentalism/2382-a-feared-force-roils-business-in-iran.

Asharq al Awsat, "Iranian Revolutionary Guard Corruption Empire," *Asharq al Awsat*, September 16, 2017.

Babak Dad's Letter To Ministry of Intelligence About Stealing Classified Documents by Ahmadinejad, news.gooya.com/2017/03/post-2162.php.

Constitution of the Islamic Republic of Iran, rc.majlis.ir/fa/content/iran_constitution.

Entekhab, Does Ahmadinejad Began his Revelations? www.entekhab.ir/fa/news/30314/منظور-احمدی-نژاد-از-برادران-قاچاقچی-و-اتصال-شبکه-هاي-قاچاق-به-صاحبان-قدرت-و-نفوذ-چه-بود.

Farhad Rezaei & Somayeh Khodaei Moshirabad, The Revolutionary Guards: From Spoilers to Accepters of the Nuclear Deal, *British Journal of Middle Eastern Studies*, Vol. 43. No. 4 2016.

Farhad Rezaei, Iran's Nuclear Program: A Study in Proliferation and Rollback (New York: Palgrave Macmillan, 2016).

FNA, "Iran Joined the Eurasian Group," *Fars News Agency*, June 12, 2016, www.farsnews.com/13950323000767.

FNA, Bonyad Mostazafan: From Establishment to Now, *Fars News Agency*, September 8, 2014, www.farsnews.com/newstext.php?nn=13930617001295.

Frederic Wehrey et al., The Rise of the Pasdaran: Assessing the Domestic Roles of Iran's Islamic Revolutionary Guards Corps, RAND, 2009.
Geneive Abdo, The Rise of the Iranian Dictatorship, *Foreign Policy*, October 7, 2009.
Graham Allison and Philip Zelikow, Essence of Decision: Explaining the Cuban Missile Crisis (Pearson, 1999).
Hashemi Rafsanjani, 'Hashemi Rafsanjani: Our Deal was for Construction, not Ownership', *Hashemirafsanjani.ir*, December 16, 2014, http://hashemirafsanjani.ir/taxonomy/term/267.
Hossein Bashirieh, A Preface to the Political Sociology of Iran of the Islamic Republic of Iran (Tehran: Negah Moaser, 2015).
Islamic Republic of Iran Constitution, rc.majlis.ir/fa/content/iran_constitution.
ISNA, Smuggling and its Eight Billion Dollars Damage to the National Economy, *Iranian Students News Agency*, www.isna.ir/news/95040107613/درنگی-در-قاچاق-کالا-و-خسارت-هشت-میلیارد-دلاری-آن-به-اقتصاد.
Jalal Al-e-Ahmad, Gharbzadegi (Tehran: Mazdâ Publishers, 1982).
Mahmoud Ahmadinejad, in meeting with IRGC Commanders, August 28, 2007.
Martin Arnold, "British Regulators Help Iranian Banks Come in From the Cold," *Financial Times*, January 31, 2016, http://www.ft.com/cms/s/0/54144ec2-c692-11e5-b3b1-7b2481276e45.html.
Ofira Seliktar, Failing the Crystal Ball Test: The Carter Administration and the Fundamentalist Revolution in Iran (USA: Praeger, 2000).
Rajanews, Gozareshi az Taasise Sepahe Pasdaran-e Enghelab-e Eslami (A Report of Establishing of Islamic Republic Revolutionary Guards), *Raja News*, April 20, 2011, http://www.rajanews.com/news/69143.
Rouhani Signs Banking Loan into Law, *Financial Tribune*, July 2, 2016, financialtribune.com/articles/economy-business-and-markets/44712/rouhani-signs-banking-reform-plan-law.
Ruhollah Khomeini, Risaleh-yi Towzih al-Masa'il [Explanation of problems] (Qom: Ruh Press, 1947).
S.K. Malik, The Quranic Concept of War (Lahore, Reprint in India: Adam Publishers and Distributors, 1979).
Stephen John Stedman, "Spoiler Problems in Peace Processes," *International Security* 22 (Fall 1997).
TAV: We trust Iran enough to carry out business there, February 10, 2004, www.hurriyetdailynews.com/tav-we-trust-iran-enough-to-carry-out-business-there.aspx?pageID=438&n=tav-we-trust-iran-enough-to-carry-out-business-there-2004-02-10.
Vali Nasr, "Meet the Decider of Tehran," *Washington Post*, December 9, 2007.

CHAPTER 2

Iran and the United States: The Rise and Fall of the Brief Détente

By creating the revolutionary paradigm in foreign policy, Ayatollah Ruhollah Khomeini positioned the Islamic Republic as the leader in the struggle against Western imperialism. The United States of America was, as noted in the preceding chapter, the immediate target of this ambitious mission. Setting the tone of the discourse, Ayatollah Khomeini called America the "Great Satan" (Shaytan-e Bozorog), which, according to the Quranic tradition, represented all the evil forces in the world under the leadership of the archdevil Iblis (the Islamic equivalent of Satan). As the "republic of virtue," it was Iran's task to liberate the world from the "Great Satan" and his coalition. For those not well versed in the Islamic theology, the Supreme Leader explained that the United States was a leader of the imperialist camp, which subjugated the "oppressed" in the developing countries. In yet another Quranic reference, Ayatollah Khomeini defined capitalism as a corrupter of the earth.[1]

Having merged a theological concept with a theory of international order, the Supreme Leader had set a very high bar for the regime realists to climb. To the hardliners, détente with the U.S. was not just a matter of a change in the direction of foreign policy but a blasphemy committed against the core mission of the Islamic Republic. Ayatollah Ahmad Jannati, head of the Assembly of Experts told the Fars news agency at the annual

[1] Ruhollah Khomeini: In this Revolution, America is the Great Satan, Sahife Imam, Vol. 10, pp. 489–497.

F. Rezaei, *Iran's Foreign Policy After the Nuclear Agreement*, Middle East Today, https://doi.org/10.1007/978-3-319-76789-5_2

rally on the last Friday of Ramadan in Iran: "We march against the arrogant powers, notably, the United States." Ayatollah Mohammad Taqi Mesbah Yazdi and his Haqqani School network took an even more radical approach. He has long criticized democracy and elections because in his opinion it came from the U.S. and held that democracy is not compatible with Islam. In July 1998, he said that "Democracy means if the people want something that is against God's will, then they should forget about God and religion. Be careful not to be deceived. Accepting Islam is not compatible with democracy." Regarding Iran's relations with the U.S., Yazdi said such relations were dangerous to the ethos of the Islamic Republic. In his words, "today America alleged that it is headman of the world, just like Pharaoh who had the same belief. The United States, as one of the most corrupted countries in the world, is trying to have cultural influence and instill its vulgar culture in Iran."[2]

But the cost of an idealistic defiance of the U.S. was too high for the country (see Chap. 1). As the de facto enforcer of the Non-Proliferation Treaty (NPT), Washington utilized the United Nations tool for rolling back proliferation. The United Nations Security Council (UNSC) Resolutions imposed sanctions banning the supply of nuclear-related technology and material. It also limited travel for many Iranian officials and froze the assets of key entities and individuals, notably Bank Saderat, Bank Melli Iran (BMI), Bank Sepah including its foreign branches, subsidiaries and a range of front companies linked with the nuclear program, or helping to develop ballistic missiles. As it turned out, the protocol for assessing Iran's threshold of economic pain was quite simple: each resolution was progressively more devastating to the economy. The resolutions further extended a freeze on the financial assets of individuals and companies associated with the nuclear project, while imposing travel restrictions on more officials involved. In what was another painful step, the Resolution banned the supply of dual-use items (civil and military) to Iran's nuclear program.[3]

[2] Rooz Online, Democracy and Human Rights have no Place in Islam, *Rooz Online*, September 1, 2010, www.roozonline.com/persian/news/newsitem/article/br-9.html; ISNA, America is our Age's Pharaoh, *Iranian Students News Agency*, September 3, 2016, http://www.isna.ir/news/95061308226/; ISNA, America Seeks to Instill Its Vulgar Culture in Iran, *Iranian Students News Agency*, October 30, 2016, http://www.isna.ir/news/95080906101/.

[3] UNSC Resolution 1737, Adopted by the Security Council at its 5612th meeting on the Non-Proliferation of Nuclear Weapons, Official Record, S/RES/1737, New York: United Nations Security Council (UNSC), 23 December 2006. UNSC Resolution 1803, Adopted

Unilaterally, Washington introduced further financial restrictions by prohibiting US banks from mediating in any capacity the transit of funds to and from Iran. As a result, several foreign banks—HSBC, Standard Chartered and Citibank—stopped dealing with Iran and urged their Iranian customers to withdraw their deposits or convert them to a different currency. The penalties for failing to do so piled up; a further squeeze was imposed on the Islamic Revolutionary Guards Corps (IRGC)-owned businesses, Iran's commercial and financial service sector, and the country's shipping industry. In addition to banning Iran from engaging in any activities related to ballistic missiles, the Resolution also imposed travel bans on individuals involved with the nuclear program, tightened the arms embargo, and froze the assets and funds of Iran's shipping lines and the IRGC. Additionally, the Resolution targeted Iran's oil supply and punished foreign groups engaged in financing the Iranian oil sector.[4]

Adding further pressure, in July 2010 President Barak Obama signed into law the Comprehensive Iran Sanctions, Accountability, and Divestment Act (CISADA). The legislation added a broad range of measures further restricting the already limited amount of U.S. trade with Iran and restricting some high technology trade with countries that allow technology that can be used in weapons of mass destruction (WMD) to reach Iran. Acting in conjunction with the UN sanctions, CISADA curtailed Iran's ability to develop its oil and gas fields. The CISADA's secondary and tertiary effects were especially damaging. In total, the CISADA made it hard to sell refined petroleum, gasoline, and gasoline production-related services, and to sell advanced equipment that would have enabled Iran to expand its own ability to produce refined petroleum. Sanctionable activities included sales of equipment with which Iran could import gasoline (such as tankers) and equipment that Iran could use to construct an energy pipeline.[5]

by the Security Council at its 5848th meeting on the Non-Proliferation of Nuclear Weapons, Official Record, S/RES/1803, New York: United Nations Security Council (UNSC), 3 March 2008.

[4] PBS, "Stuck Between America and Iran," *PBS, Tehran Bureau*, September 18, 2009, http://www.pbs.org/wgbh/pages/frontline/tehranbureau/2009/09/stuck-between-america-and-iran.html, UNSC Resolution 1929, Adopted by the Security Council at its 6335th meeting on the Non-Proliferation of Nuclear Weapons, Official Record, S/RES/1929, New York: United Nations Security Council (UNSC), 9 June 2010.

[5] CISADA, "Comprehensive Iran Sanctions, Accountability, and Divestment Act of 2010 (CISADA)," *The Department of the Treasury*, July 1, 2010, http://www.treasury.gov/resource-center/sanctions/Documents/hr2194.pdf, Kenneth Katzman, Iran sanctions (Washington: Congressional Research Service, 2011).

Washington unilaterally beefed-up sanctions against those individuals and entities investing or supporting the development of Iran's oil sector. In December 2011, the assets of financial institutions trading with the Central Bank of Iran (CBI) in the oil sector were frozen. On January 23, 2012, the EU imposed new sanctions, including a full ban on Iranian oil exports and freezing the assets of BMI and CBI. Two weeks later, on February 6, 2012, the U.S. Executive Order (EO) 13,599 imposed a set of sanctions on CBI and other financial institutions and seized the assets of CBI in the United States.[6]

Most critically, on March 15, acting on the EU order, the Belgian-based Society for Worldwide Interbank Financial Telecommunication (SWIFT) declared it would bar Iranian banks from its network. The involvement of SWIFT, the world's largest electronic payment system, represented a new phase of sanction statecraft—turning smart sanctions into super smart ones. The SWIFT declaration coincides with reports that major currency exchangers in the United Arab Emirates (UAE) had stopped handling the Iranian Rial, a development that further reduced Tehran's ability to trade and acquire hard currency. The financial disruption proved devastating to Iran's economy in many ways. CBI and other major Iranian banks—Bank Mellat, Bank Tejarat, Future Bank, Bank Refah, Post Bank, Persia International Bank and Europäisch-Iranische Handelsbank—could not conduct international transactions. Virtually overnight Iran became financially isolated, as money could not flow in or out of the country through official banking channels.[7]

[6]Council Regulation, "Council Regulation (EU) No 267/2012 of 23 March 2012 Concerning Restrictive Measures Against Iran and Repealing Regulation (EU) No 961/2010," *Official Journal of the European Union* 23 March 2012, http://eur-lex.europa.eu/LexUriServ/LexUriServ.do?uri=OJ:L:2012:088:0001:0112:en:PDF.

UNSC. Report, "Chronology of Events: Iran," *Security Council Report.org*, December 1, 2014, http://www.securitycouncilreport.org/chronology/iran.php?page=1, Executive Order 13599, Executive Order 13599 of February 5, 2012: Blocking Property of the Government of Iran and Iranian Financial Institutions, Washington, DC: Presidential Documents, Vol. 77, 8 February 2012.

[7]Rick Gladstone and Stephen Castle, Global Network Expels as Many as 30 of Iran's Banks in Move to Isolate Its Economy, *The New York Times*, March 15, 2012; BBC, "Iran's banks to be blocked from global banking system," *BBC News*, March 15, 2012, http://www.bbc.com/news/business-17390456; Rick Gladstone and Stephen Castle, Global Network Expels as Many as 30 of Iran's Banks in Move to Isolate Its Economy, Brendan Taylor, "Chapter Three: Sanctioning Iran," *The Adelphi Papers*, Vol. 49, No. 411 (2009): pp. 59–100.

Financial sanctions made receiving payment for oil, a vital component of Iran's economy and a key source of government income and foreign currency, substantially more difficult. They limited Iran's ability to conduct financial transactions and finance trade or, at best, increased their costs dramatically. For instance, according to reports by global financial organizations, many foreign banks, and financial institutions were reluctant to process transactions for Iranian citizens and businesses, even when it was not clear that these transactions would trigger sanctions. Unable to arrange the financing for their foreign trade, company officials were forced to transfer suitcases of cash to shady foreign banks using the services of street-level money changers. With brokers demanding fees every step of the way, this practice was not only costly but also risky, as cash was a tempting target for thieves.[8]

Not satisfied with SWIFT alone, the United States pressured Iran's top oil customers—Japan, South Korea, and India—to cut their imports. In case of noncompliance, Asian companies involved in oil trade faced an array of penalties—being barred from receiving U.S. Export–Import Bank financing, U.S. export licenses, and loans over $10 million from the United States financial institutions, among others. In fact, Washington forced these countries to choose between doing business with the United States or Iran. Additional US sanctions tightened Iran's economic noose. As of May 1, 2012, EO 13608 targeted persons engaged in misleading practices to withhold or obscure information about Iranian links to financial transactions.[9]

Acting under duress, Ayatollah Khamenei accepted the Joint Comprehensive Plan of Action (JCPOA), known in Iran as Barname Jame Eghdam Moshtarak (BARJAM), a decision that foreign policy experts viewed as a tipping point in the decades-long struggle between the hardline ideologues and the normalizers. But even Khamenei's stamp of approval did not save the latter from rhetorical abuse. Ayatollah Jannati implied that Rouhani was a "traitor" and he added that "some people want to betray

[8] GAO, U.S. and International Sanctions Have Adversely Affected the Iranian Economy: Report to the Chairman, Committee on Homeland Security and Governmental Affairs, U.S. Senate, Washington, DC, GAO-13-326, February 2013, Thomas Erdbrink, Iran Staggers as Sanctions Hit Economy, *The New York Times*, September 30, 2013.

[9] Executive Order 13608, Executive Order 13608: Prohibiting Certain Transactions With and Suspending Entry Into the United States of Foreign Sanctions Evaders With Respect to Iran and Syria, Presidential Documents, Washington, DC: U.S. Government Publishing Office, Vol. 77, 1 May 2012.

the nation, and they think that the enemy [the United States] will fulfill its promises." The Principalists had also disparaged the terms of the JCPOA, adding that the normalizers "sold the country to the foreigners."[10]

Ayatollah Khamenei warned that the negotiations were all about the nuclear issue and nothing else. "We will not negotiate with the U.S. on domestic, regional and arms issues at all because U.S. objectives in the region opposes our regional objectives, and since the U.S. policy in the region is based on stirring insecurity and confrontation with the regional states and Islamic Awakening which runs counter to the pivotal policies of the Islamic Republic of Iran."[11]

Rhetoric aside, the expected détente with Washington mandated Iran to adhere to a rigid four-point plan of normalization. All four dimensions of the plan have required cooperation from the parastatals or, at the very least, a commitment to abstain from spoiling actions. But the Rouhani government could also count on the Obama administration, which was eager to prove that the nuclear deal was just a prelude to a broader détente with the regime. All this changed when Donald J. Trump entered the White House in January 2017.

As a candidate, Trump described the JCPOA as a "disaster" and "the worst deal ever negotiated", which could lead to a "nuclear holocaust," and adding that he would "dismantle" it. Secretary of Defense Major General James Mattis developed a dim view of Iran while serving as the head of Central Command (CENTCOM) under the Obama administration. Mattis called for cross-border actions against Iran, only to be overruled by President Obama, which led to Mattis's resignation. Lieutenant General R.H. McMaster listed Iran as one of the threats to the U.S., which "has been fighting a proxy war against the U.S. since 1979," and "building militias" beyond the control of Middle Eastern governments to use as leverage against them. McMaster pointed to "isolated Iranian actions" becoming "part of a geopolitical realignment that cuts against U.S. interests." He cautioned: "Iran is on the move…intimidating allies, and using

[10] FNA, Negotiation with the West is Betrayal, *Fars News Agency*, October 28, 2016, www.farsnews.com/13950708000252.

[11] TNA, Iran's Supreme Leader Emphasized that Tehran will Negotiate on Nuclear Issue Only, *Tasnim News Agency*, March 21, 2015, https://www.tasnimnews.com/fa/news/1394/01/01/692344/رهبر-ایران-تاکید-کرد-که-تهران-تنها-درباره-موضوع-هسته-ای-با-آمریکا-مذاکره-می-کند; Kayhan Newspaper, Ayatollah Khamenei Called America "Great Satan," *Kayhan Newspaper*, June 6, 2015, kayhan.ir/fa/news/46577.

proxy armies and unconventional forces to challenge the post-World War II political order."[12]

Rex Tillerson, the former Secretary of State, has repeatedly described Iran as a highly destabilizing force in the Middle East. During a press conference at the State Department, Tillerson said "Iran is the world's leading state sponsor of terrorism, and is responsible for intensifying multiple conflicts and undermining U.S. interests in countries such as Syria, Yemen, Iraq and Lebanon, and continuing to support attacks against Israel. An unchecked Iran has the potential to travel the same path as North Korea, and take the world along with it."[13]

The Republican-dominated Congress added its list of concerns. By all accounts, both the executive and the legislative branch were ready to apply increased scrutiny to Iran's foreign policy starting with the nuclear deal.

Implementing the JCPOA: The Making or Breaking of Normalization

The JCPOA, which took more than two years to negotiate, was a dense technical document. In a nutshell, the Iranians agreed to cut the number of centrifuges to 6014, but only 5060 of the first-generation IR-1 models were approved to operate for the next 10 years. The IR-2 m and other advanced models were to be decommissioned and stored in Natanz under the IAEA safeguards. The Fordow facility was ordered to cease uranium enrichment and research into uranium enrichment for 15 years. Once converted into a nuclear, physics, and technology center, Fordow could house no more than 1044 IR-1 centrifuges in six cascades in one of its wings. Two of the cascades were to be modified to produce radioscopes for medical, agricultural, industrial, and scientific use. The other four cascades were to remain idle. Iran could keep 300 kg uranium enriched to 3.67 percent and would reduce its stockpile of 10,000 of low-enriched uranium (LEU) either by blending it down or selling it abroad.[14]

[12] Reuters, Trump election puts Iran nuclear deal on shaky ground, *Reuters*, November 9, 2016, www.reuters.com/article/us-usa-election-trump-iran-idUSKBN13427E; H.R. McMaster, Probing for Weakness, *The Wall Street Journal*, March 23, 2016, https://www.wsj.com/articles/probing-for-weakness-1458775212.

[13] TWP, Tillerson's full statement on Iran, *The Washington Post*, April 19, 2017, https://www.washingtonpost.com/video/national/watch-tillersons-full-statement-on-iran/2017/04/19/4a1bc8ba-2540-11e7-928e-3624539060e8_video.html.

[14] JCPOA, "Joint Comprehensive Plan of Action," 14 July 2015, https://www.justsecurity.org/wp-content/uploads/2015/07/271545626-Iran-Deal-Text.pdf.

The Arak heavy water research reactor was ordered to convert with the help of "Working Group" E3/EU+3 (the P5+1 and possibly other countries) to support peaceful nuclear research and production needs. The power of the redesigned reactor was not to exceed 20 MW, and it was blocked from producing weapon-grade plutonium. Some of the heavy water from the Heavy Water Power Plant (HWPP) was to be used for modernizing the reactor and the rest exported to international markets. The agreement encouraged Iran to follow new technology, which favored light water reactors and made it impossible to accumulate heavy water or build heavy water reactors for 15 years.[15]

According to the JCPOA's architects, the set of provisions was designed to lengthen Iran's breakout capacity. Breakout capacity refers to the time it would take Iran to produce highly enriched uranium for one weapon should the regime decide to leave the NPT. Indeed, drastically limiting the number and quality of the centrifuges and the uranium stockpile was said to lengthen the breakout period from two months in 2013 to one year for the next ten years. The one-year timetable pertains to uranium production and does not include predictions about nuclear weaponization. The latter comprises two steps. First, a nuclear payload needs to be fabricated and fitted on a missile. Miniaturizing a payload is a relatively difficult task and in the case of Iran the estimate for a working weapon, known as "effective breakout time," is at least one year.[16]

To prevent Iran from cheating, the JCPOA offered a strict safeguards protocol based on electronic monitoring, visits by IAEA inspectors, and unspecified cyber sleuthing. In addition to the IAEA, national intelligence agencies of the United States, Israel, and other countries monitor Iran's program unofficially.[17]

To this end, IAEA developed a multilayered oversight "over Iran's entire nuclear supply chain, from uranium mills to its procurement of nuclear-related technologies." In Fordow and Natanz a "round-the-clock access"

[15] Ibid.

[16] JCPOA, "Joint Comprehensive Plan of Action," 14 July 2015, https://www.justsecurity.org/wp-content/uploads/2015/07/271545626-Iran-Deal-Text.pdf; Ishaan Tharoor, "The historic nuclear deal with Iran: How it works," *The Washington Post*, 14 July 2015, https://www.washingtonpost.com/news/worldviews/wp/2015/07/14/the-historic-nuclear-deal-with-iran-how-it-works/.

[17] David Sanger, & Broad, In Iran Talks, U.S. Seeks to Prevent a Covert Weapon, *The New York Times*, November 22, 2014; Moshe Ya'alon, Current Iran framework will make war more likely, *The Washington Post*, April 8, 2015.

would be available, including continuous monitoring via surveillance equipment. The IAEA would use a new generation of monitoring technology. The list includes fiber-optic seals on equipment that electronically send information to the IAEA and infrared satellite imagery to detect covert sites. Moreover, "environmental sensors that can detect minute signs of nuclear particles," tamper-resistant, radiation-resistant cameras, computerized accounting programs for information gathering and anomalies detection through the use of big data, which monitors Iran's dual-use imports are available and are particularly promising. Human monitoring has also intensified, as the number of IAEA inspectors tripled from 50 to 150.[18]

A particular procedure addressed IAEA's concerns about Iran's efforts to develop nuclear capabilities in non-declared sites. The IAEA has the right to access such sites to verify the absence of undeclared nuclear materials and activities or activities inconsistent with the agreement. Should Iran refuse such a request or otherwise fail to satisfy the IAEA's concerns, the JCPOA sets in motion a special process of arbitration lasting up to 24 days. Iran and the IAEA have 14 days to resolve disagreements among themselves.[19]

Should they fail, the Joint Commission—a body composed of representatives from the United States, Great Britain, France, Germany, Russia, China, EU—and Iran would have one week in which to consider the intelligence that initiated the IAEA request. A majority of the Commission (at least five of the eight members) could then decide to require Iran to act. The majority rule provision means the United States and its European allies could not be vetoed by Iran, Russia, and China. If Iran did not comply with the decision within three days, sanctions would be automatically imposed under the "snapback provision."[20]

[18] JCPOA, "Joint Comprehensive Plan of Action," 14 July 2015, https://www.justsecurity.org/wp-content/uploads/2015/07/271545626-Iran-Deal-Text.pdf; Ishaan Tharoor, "The historic nuclear deal with Iran: How it works," *The Washington Post*, 14 July 2015, https://www.washingtonpost.com/news/worldviews/wp/2015/07/14/the-historic-nuclear-deal-with-iran-how-it-works/.

[19] Heroux, Is Iran's Nuclear Program Really Getting a 24-Day Pass? Huffington Post, August 6, 2015. Hersh, "The coming wars," *The New Yorker*, 2005, http://www.newyorker.com/magazine/2005/01/24/the-coming-wars; Tower, "Deal Will Give Iran 24 Days to Cover Up Suspected Nuclear Sites," *TheTower.org*, July 14, 2015, http://www.thetower.org/2296-deal-will-give-iran-24-days-to-cover-up-suspected-nuclear-sites/.

[20] JCPOA, "Joint Comprehensive Plan of Action," 14 July 2015, https://www.justsecurity.org/wp-content/uploads/2015/07/271545626-Iran-Deal-Text.pdf.

The JCPOA mandated two additional steps. Iran signed a separate agreement with the IAEA to investigate past experimentation with Possible Military Dimension (PMD) applications, a demand that the Agency has insisted on for a good part of the last decade. Iran committed to signing the Additional Protocol to the NPT, which gives the IAEA a broader verification mandate than the original treaty.[21]

In return for the restrictions, Iran obtained sanction relief. The agreement promised not to impose UN or EU nuclear-related sanctions or restrictive measures after the UNSC, the United States and the European Union had lifted the previous sanctions. Iran had to take action to meet its commitments before the agreement was implemented on January 16, 2016. Sanction relief was conditional: should Iran renege on its commitments, they would be "snapped back," that is imposed again.[22]

Facing fierce resistance from the Iranians over the inclusion of the ballistic missiles program in the JCPOA, the P5+1 agreed to leave the topic in the hands of the UNSC. In a further compromise, Iranian officials sought to soften the language of UNSC Resolution 1929 (2010), which stipulated: "Iran shall not undertake any activity related to ballistic missiles capable of delivering nuclear weapons." Passed on July 20, 2015, the new UNSC Resolution 2231endorsed the nuclear deal and was more permissive: "Iran is called upon not to undertake any activity related to ballistic missiles designed to be capable of delivering nuclear weapons."[23]

Missile tests are essential to Iran's large and advanced program, which boasts thousands of short- and medium-range missiles, some 300 intermediate-range missiles, and even a space program. Eager to explore the boundaries of Resolution 2231, the IRGC Aerospace Force (IRGCAF) launched several ballistic missiles that could potentially carry a nuclear payload. Among them were the mid-range Qhadr-H and Qhadr-F with an estimated range of 2000 km, the Emad with a range of 1700 km, the Zolfiqar with a range of 750 km, the Qiam-1 with an estimated range of 800 to 1000 km, and the Khorramshahr with an estimated range of 2500 to 4000 km. Deputy Chief of Staff of the Iranian Armed Forces, Brigadier

[21] TCACNP, "Factsheet: Iran and the Additional Protocol," *The Center for Arms Control and Non-proliferation*, July 14, 2015, armscontrolcenter.org/factsheet-iran-and-the-additional-protocol.

[22] JCPOA, "Joint Comprehensive Plan of Action," 14 July 2015, https://www.justsecurity.org/wp-content/uploads/2015/07/271545626-Iran-Deal-Text.pdf.

[23] UNSC Resolution 2231 (2015), Adopted by the Security Council at its 7488th meeting, on 20 July 2015, United Nations Security Council, S/RES/2231 (2015).

General Ali Abdullahi, explained, "We test-fired missiles with a range of 2000 kilometers and a margin of error of eight meters, which can easily reach Israel."[24]

In March 2016, a spokesman for the Obama administration stated that the launches might have violated Resolution 2231. Under pressure from Congress, the White House imposed new sanctions on the Revolutionary Guards, but Brigadier General Ali Hajizade, the commander of the Aerospace Force, declared that sanctions would not deter Iran from testing ballistic missiles. While President Rouhani's government was not keen on a confrontation with the international community, the Foreign Ministry was forced to side with the Revolutionary Guards. Nonetheless, Rouhani did complain to the Supreme Leader when the Guards paraded a missile decorated with the Hebrew slogan "Israel should be wiped off the map."[25]

Provocations aside, the Revolutionary Guards justified their launch by drawing on a legal loophole that Resolution 2231 had created in leaving the definitions of a "nuclear-capable missile" somewhat vague. Since there is no clear consensus on the subject, UNSC relies on an appointed Panel of Experts to study the relevant ballistic missiles tests. Upon receiving the Panel's report, UNSC declares whether the missile is nuclear-capable. The range is a primary consideration and, per the Missile Technology Control Regime (MTCR), an informal group that assesses nuclear-capable missile technology, a missile with a range of 300 km and a payload of 500 kg is considered "nuclear-capable."[26]

Days after Donald Trump took office, on January 29, 2017, the Aerospace Force tested a new missile, Khorramshahr, in the Semnan launch site, about 140 miles east of Tehran. The U.S. intelligence community suggested that the Khorramshahr missile was a variant of the highly debated intermediate-range North Korean Musudan missile, with an estimated range of 4000 km. The range of the Iranian missile, however, was not clear; the weapon exploded upon reentry after a 980-kilometer flight. Still, Washington reacted immediately. On February 3, 2017, the Treasury Department placed new sanctions on 25 individuals and entities involved in developing the missile. Shortly after, Michael T. Flynn, the then national security adviser, stated that Iran "was put on notice," a threat which left

[24] TCACNP, "Factsheet: Iran and the Additional Protocol," *The Center for Arms Control and Non-proliferation*, July 14, 2015, armscontrolcenter.org/factsheet-iran-and-the-additional-protocol.

[25] Ibid.

[26] Ibid.

little doubt that the United States would react more severely should Iran test another missile in violation of Resolution 2231.[27]

Tehran's response at the time indicated a certain tension between the hardline Principalists and the Rouhani government. Representing the former, the Defense Minister Hossein Dehghan claimed such launches would continue with "enhanced speed and precision." The *Kayhan* newspaper considered to be the mouthpiece of the Supreme Leader described the test as a "historical turn." Members of the Abadgaran (Principalist) faction in the Majlis proposed to impose sanctions on 15 American companies that collaborated with Israel for alleged violations of human rights. The measure would have been largely symbolic because none of the companies had business with Iran. The moderates, however, urged caution. The government-run *Iran Daily* wrote: "This measure by Iran provides an excuse for Trump to take actions against Iran, increasing his intention to disrupt the status quo resulting from the Iran nuclear deal."[28]

For all intents and purposes, the moderates won the day. After being "put on notice," the Guards' Aerospace Force carried one test of the short-range surface-to-air Mersad on February 8, 2017. Despite Rouhani's subsequent proclamation that Iran does not require permission from the United States, the Mersad was well within the limits of Resolution 2231. The IRGC did not attempt to test other missiles deemed nuclear-capable; a self-imposed restraint that was even more remarkable because it favored Rouhani in the May 19 presidential election. Persistent rumors in intelligence circles indicated that the hardliners accepted Rouhani's suggestion not to test President Trump whom the Ministry of Intelligence and Security (MOIS) described as dangerous and unpredictable. If anything, the White House decision to launch a Tomahawk strike in Syria and the dropping of the monstrous GBU-43/Massive Ordnance Air Blast (MOAB)

[27] Karen DeYoung, Trump administration says it's putting Iran 'on notice' following missile test, February 1, 2017, https://www.washingtonpost.com/world/national-security/2017/02/01/fc5ce3d2-e8b0-11e6-80c2-30e57e57e05d_story.html?utm_term=.b1a66e19d1a2.

[28] TNA, Defense Minister, The Recent Missile Test was Successful, February 1, 2017, *Tasnim News Agency*, https://www.tasnimnews.com/fa/news/1395/11/14/1315991/وزیر-دفاع-آزمایش-موشکی-اخیر-موفقیت-آمیز-بود; Kayhan Newspaper, Historical Turn and Uts Requirements, *Kayhan Newspaper*, February 1, 2017, kayhan.ir/fa/mobile/news/96856/927; Iran Newspaper, the Consequences of Iran's Missile Test, *Iran Newspaper*, February 2, 2017, www.iran-newspaper.com/Newspaper/BlockPrint/169767.

on an ISIS stronghold in Afghanistan reinforced Tehran's anxieties. In a rather uncharacteristically low-key response, Ayatollah Khamenei noted, "the Americans' action is a strategic mistake, as they are repeating the mistakes of their predecessors."[29]

Iran scored an important endorsement when the IAEA released its report on February 24, 2017. The Agency concluded that Iran fulfilled its nuclear-related commitments under the JCPOA, and on matters related to verification and monitoring in Iran in light of the UN resolution 2231, the Agency has verified and monitored Iran's implementation of its nuclear-related commitments under the JCPOA. The reports indicate that Iran has fulfilled its JCPOA commitments and did not continue its nuclear activities. In other words, Iran addressed major nuclear concerns. It stopped the construction of the existing Arak heavy water research reactor (IR-40 Reactor) and it also ceased producing or testing natural uranium pellets, fuel pins or fuel assemblies specifically designed for the support of the IR-40 Reactor. All existing natural uranium pellets and fuel assemblies have remained in storage under continuous agency monitoring.[30]

The reports indicate that Iran has not carried out activities related to reprocessing at its nuclear sites, nor it has operated any of its declared facilities for reconverting fuel plates or scrap into UF6 and has not built any new facilities for such a purpose. In other words, the agency certified that Iran has been transparent enough with regards to its nuclear activities allowed by the JCPOA.[31]

The IAEA reports have played a part in the Trump administration's decisions on the JCPOA. Under the Iran Nuclear Agreement Review Act (INARA) of 2015, the White House is required to review Iran' s compliance every three months and notify Congress so it can certify Iran officially. Based on the IAEA report and information from the intelligence service, on April 18, 2017, Secretary Tillerson certified Iran as compliant. The letter stated that "the conditions of Section 135(d) (6) of the Atomic Energy Act of 1954 (AEA), as amended, including as amended by the Iran

[29] Farhad Rezaei, Iran's Ballistic Missiles Program: Changing Course in the Trump Era? *Harvard Belfer Center, Iran Matters*, May 08, 2017, http://www.belfercenter.org/publication/irans-ballistic-missiles-program-changing-course-trump-era.

[30] IAEA Report by the Director General, Verification and monitoring in the Islamic Republic of Iran in light of United Nations Security Council resolution 2231 (2015), GOV/2017/10 February 24, 2017.

[31] Ibid.

Nuclear Agreement Review Act of 2015 (Public Law 114–117), enacted May 22, 2015, are met as of April 18, 2017."[32]

The administration certified Iran again on June 2, 2017, and January 9, 2018, but the president made his reluctance to do so well known. He signaled he might refuse to declare Iran compliant during the next certification call on May 12, 2018. The president instructed members of his administration to search for evidence of Iran's violation of the agreement and loopholes that make violations possible. Although the administration did not point to specific violations, the White House refused to certify Iran's compliance because the deal did not fit American national interests, a stipulation allowed by the INARA. According to the bill, the Congress would have to determine by December 15, 2017 whether to reimpose the nuclear sanctions.

Regardless of the future of the JCPOA, the Trump White House moved to declare the IRGC-QF as a terror organization and ordered Treasury to impose a serious of sanctions on the organization. Because of the extensive economic holdings of the Guards, the new sanctions are expected to undermine the still fragile Iranian economy, a fact that the normalizers were quick to point out.

Limiting Terror and Military Interventions: Mission Impossible?

By the time Rouhani came to power in 2013, the IRGC-QF terror activities had generated high-level condemnations, as pointed out in the previous chapter. Excluded from the decision-making of the hardline circles, the president, and his minister could do little to limit the terror and the destabilization, which the Guards had engaged in as part of Khomeini's mission. According to the 2016 report, they had "remained the foremost state sponsor of terrorism in 2015, providing a range of support, including financial, training and equipment, to groups around the world."[33]

Remarkably, during the 2017 campaign, Rouhani and his followers tried to start a public debate on the cost of exporting the revolution. On April 29, 2017, at a campaign event in Esfahan province, Gholam Hossein

[32] Rex W. Tillerson, Trump Administration Undergoing Interagency Review of Iran Deal, April 18, 2017, https://www.state.gov/secretary/remarks/2017/04/270315.htm.

[33] USDS, Strategic Assessment Bureau of Counterterrorism and Countering Violent Extremism, Country Reports on Terrorism 2015, Chap. 1, www.state.gov/j/ct/rls/crt/2015/257513.htm.

Karbaschi, former Mayor of Tehran and head of the Reformist Executives of Construction Party (Hezb-e Kaargozaaraan-e Saazandegi) criticized Iran's military presence in the region. He stated, "We too want peace in Syria, Lebanon and Yemen, for the oppressed to be defended and the Shiites to be supported. But this cannot be done by giving money, buying arms and killing."[34]

Unlike in the case of JCPOA, however, the Rouhani government had little leverage over the parastatals, which oversaw revolutionary export. To complicate matters, the Arab Spring of 2011 inspired Ayatollah Khamenei to increase the regime's campaign of destabilization. Using the same theological–international relations reasoning as his predecessor, Khamenei concluded that the Arab Spring was a God-given opportunity for Iran to finish its revolutionary project. In his view, "exporting revolutionary culture to other countries promotes Islamization of Muslim societies," a task made easier by the political vacuum created in the region. He credited the loss of control by Americans and their secular allies for the growth of Islamist groups such as Islamic State of Iraq and Syria (ISIS).[35]

Ayatollah Khamenei has also supported the IRGC's plan to limit American power-projection naval operations in the Persian Gulf and beyond as well as create disruption in select maritime choke points. Admiral Habibollah Sayyari, the commander of the Navy, revealed that the Supreme Leader had urged the naval forces to secure Iran's maritime prominence and influence. Translating Khomeini's directive, the Iranian navy developed a 20-year plan known as the Strategic Triangle Maritime, Irregular Warfare (MIW). The ideas were derived from the lessons of Iran's efforts to mine the Straits of Hurmuz during the Tanker War in 1987. After suffering damage to one of its ships, the US launched Operation Praying Mantis, which destroyed several Iranian offshore oil facilities.

It was, thus, MIW's top priority to slow down the American navy's access to the Straits of Hormuz and hamper its power-projecting naval operations elsewhere in the region and beyond. If successful, the plan would create a Maritime Triangle extending from the Straits of Hormuz to Bab al-Mandeb in the Red Sea, and the Straits of Malacca in the Indian Ocean. Moreover, the Maritime Triangle would give Iran an opportunity

[34] Karbaschi Criticizes Iran's Military Presence in the Syrian War, *YouTube*, April 29, 2017, https://www.youtube.com/watch?v=wYW0qr2pHf0.

[35] Yvette Hovsepian-Bearce, The Political Ideology of Ayatollah Khamenei. Out of the Mouth of the Supreme Leader (London: Routledge, 2016), 353.

to disrupt some of the most heavily trafficked commerce and oil trade choke points like Bab al-Mandeb and the Persian Gulf. The Maritime Triangle plan strives to increase the cost of American naval operations by employing an asymmetrical hybrid naval approach subsumed in the so-called Anti-Access/Anti-Denial (A2/AD) strategy.

Some of the A2/AD capabilities are traditional, such as land- and sea-based anti-ship cruise missiles, which are under the purview of the regular (blue-water) Islamic Republic of Iran Navy (IRIN). Others are less conventional such as the use of naval mines, which small commercial vessels can deploy covertly. The plan seeks to use small, fast-attack craft best suited for swarming operations or suicide attacks, such as those used against the USS *Cole*. The Revolutionary Guards' Neyroye Daryaee Sepah-e Iran (NEDSA), the Sepah Navy Special Force has also used naval drones. The NEDSA, under Rear Admiral Ali Fadlavi is in charge of the unconventional capabilities.

Since July 2015, NEDSA has been involved in nine incidents, most of which were against American vessels patrolling the Gulf. Reportedly, the Revolutionary Guards have 3000 to 5000 speedboats of several types which have a speed of over 110 knots (203 km/h) and are equipped with heavy machine guns and 107 mm or other similar rocket launchers and missiles, including Zafar, Nasr, Noor, and Ghader. These are radar-guided anti-ship cruise missiles capable of destroying 1500-ton targets and damaging even larger ones.[36]

Over the past few years, NEDSA carried out numerous swarming attacks against American ships in the Persian Gulf and Strait of Hormuz. On December 26, 2015, the Guards' military vessel fired several unguided rockets toward U.S. and French military vessels and commercial craft in international waters. Two weeks later, on January 12, 2016, an unarmed Guards' drone flew toward both the French aircraft carrier *Charles de Gaulle* and the USS *Harry S. Truman* while they were in the international waters of the Persian Gulf. The Guards' forces also seized two U.S. Navy vessels and held them on Iran's Farsi Island in the Persian Gulf. Nine men and one woman were aboard the small riverine boats. The sailors were safely released to the U.S. Navy after an overnight flurry of diplomatic efforts.[37]

[36] An Overview of the IRGC Speedboats, *Mashregh News*, https://goo.gl/rGiDx9.

[37] CNN, See Iranian rocket fly by U.S. carrier edition.cnn.com/videos/us/2016/01/09/iran-rocket-us-ship-video-orig-vstan.cnn; CNN, Iranian surveillance drone flies over U.S. aircraft

On July 12, 2016, five Guards' vessels, including four speedboats and a guided missile patrol ship, came as close as 500 meters to a U.S. warship hosting a top general in the Strait of Hormuz. On August 23, 2016, the guided-missile destroyer USS *Nitze* (DDG-94) was followed and harassed by four patrol boats of the Guards in waters of the Persian Gulf. The destroyer was near the Strait of Hormuz when high-speed patrol boats came at the ship without responding to hails, or warning flares fired from *Nitze*, according to a video of the incident provided to USNI News.[38]

On August 25, 2016, three more U.S. ships were harassed by Iranian patrol boats in three separate incidents, according to U.S. 5th Fleet spokesman Cmdr. One of the encounters resulted in a U.S. ship firing a warning shot at a Guards' patrol boat with a 0.50 caliber machine gun to ward off a dangerous approach, according to the service.[39] On September 4, 2016, seven Iranian boats harassed a U.S. Navy ship in the Persian Gulf. Reportedly, the U.S. Navy ship narrowly avoided colliding with one. Three of the fast-inshore attack craft (FIAC) maneuvered close to USS *Firebolt*, mirroring the ship's course and speed at a distance of about 500 yards for about eight minutes before leaving. Another FIAC then sped towards *Firebolt* and stopped directly in front of the ship, causing the *Firebolt* crew to maneuver to avoid a collision.[40]

On September 13, 2016, Revolutionary Guards threatened to shoot down a U.S. Navy surveillance aircraft flying close to Iranian territory in the Persian Gulf that it said was encroaching on its airspace. On October 10, 2016, U.S. warships were targeted by rebels from the territory controlled by Iran-aligned Houthi rebels in Yemen. The missile attack on the USS *Mason* failed, and both missiles impacted the water before reaching the ship. On November 28, 2016, a Guards' small boat trained a machine

carrier in Persian Gulf, January 29, 2016, edition.cnn.com/2016/01/29/politics/iran-drone-uss-harry-truman/?utm_source=Sailthru&utm_medium=email&utm_campaign=New%20Campaign&utm_term=%2ASituation%20Report; U.S. Sailor Flap: Model for Diplomacy? January 12, 2016, iranprimer.usip.org/blog/2016/jan/12/10-us-sailors-held-iran-be-returned.

[38] Reuters, After nuclear deal, Iran Sticks to Wary Tactics in Key Gulf Waterway, July 12, 2016, www.reuters.com/article/us-usa-iran-warship-idUSKCN0ZR2E0; Reuters, Iran's vessels make 'high-speed intercept' of U.S. ship: U.S. official August 25, 2016, www.reuters.com/article/us-usa-iran-military-idUSKCN10Z2OP.

[39] USNI, Three More U.S. Navy Ships Harassed by Iranian Patrol Boats August 25, 2016, https://news.usni.org/2016/08/25/three-u-s-navy-ships-harassed-iranians-patrol-boats.

[40] USNI, Iranian Boats Harass Another U.S. Navy Patrol Coastal Ship In Persian Gulf, September 6, 2016, https://news.usni.org/2016/09/06/iranian-boats-harass-another-u-s-navy-patrol-coastal-ship-persian-gulf.

gun on a U.S. Navy helicopter flying off an American aircraft carrier in the Strait of Hormuz. A Navy official called the Guards' behavior "unsafe and unprofessional," and an "escalating measure." On January 9, 2017, USS *Mahan* fired warning shots at four vessels of the Guards near the Strait of Hormuz. In that incident, the *Mahan* tried, via bridge-to-bridge radio communication, to order the boats to stop, but the vessels didn't respond, prompting the destroyer to fire three warning shots with a 0.50 caliber machine gun. On April 25, 2017, a U.S. Navy destroyer had another close encounter with an Iranian Revolutionary Guard "fast attack craft" in the Persian Gulf. Two U.S. officials told Fox News that the Iranian ship came within 1000 yards of the guided-missile destroyer USS *Mahan* with its weapons manned.[41]

Recalling that the negotiated political order grants the Rouhani administration little influence over the parastatal Revolutionary Guards, harassment of American vessels has been an effective way to tarnish the normalization project. Not surprisingly, Ayatollah Khamenei, who accepted the nuclear deal under economic duress, seemed to have blessed the NEDSA initiative. On October 7, 2015, Khamenei praised the Revolutionary Guards' naval units for their "vigilance" in the Persian Gulf. The recent encounters have brought more commendation from the Supreme Leader and the media outlets that support him. In January 2016, the Revolutionary Guards seized two U.S. Navy boats and ten sailors after they entered Iranian territorial waters. Khamenei praised the move saying that they did "God's deed," and subsequently gave medals to the commanders in charge of the capture.[42]

[41] Fox News, Iran threatened to shoot down U.S. Navy spy planes in the Persian Gulf, September 13, 2016, www.foxnews.com/world/2016/09/13/iran-threatened-to-shoot-down-us-navy-spy-planes-in-Persian-gulf.html; Reuters, U.S. Navy ship targeted in failed missile attack from Yemen: U.S., October 10, 2016, www.reuters.com/article/us-yemen-security-usa-ship-idUSKCN12A082?il=0; Fox News, Iranian small boat pointed weapon at U.S. Navy Helicopter in Strait of Hormuz, officials say, November 28, 2016, www.foxnews.com/world/2016/11/28/iranian-small-boat-pointed-weapon-at-us-navy-helicopter-in-strait-hormuz-officials-say.html; Fox News, A Navy destroyer fires warning shots at Iranian vessels in Strait of Hormuz, January 09, 2017, www.foxnews.com/world/2017/01/09/navy-destroyer-reportedly-fires-warning-shots-at-iranian-vessels-in-strait-hormuz.html; Fox News, Navy destroyer has close encounter with Iran vessel in the Persian Gulf, April 25, 2017, www.foxnews.com/world/2017/04/25/navy-destroyer-has-close-encounter-with-iran-vessel-in-persian-gulf.html.

[42] Awarding Fath Medal to the Commander who Arrested American Invaders," *Ayatollah Khamenei's Website*, September 26, 2016, farsi.khamenei.ir/news-content?id=32,124; God's

Since the Trump administration had put Iran "on notice" on January 29, 2017, there was only one swarming incident, on March 6, 2017, in which reportedly multiple fast-attack vessels from the Guard came close to a U.S. Navy ship, the USNS *Invincible*, in the Strait of Hormuz, forcing it to alter course. Still, as General Joseph Votel, head of the U.S. military's Central Command (CENTCOM) noted, "in a relatively compressed space here, there is a great opportunity for miscalculations." The official said attempts were made to establish radio contact, but there was no response from the Guards' vessels, which was deemed "unsafe and unprofessional" but an improvement in Iran's behavior compared to its previous encounters. The U.S. Department of Defense spokesman Captain Jeff Davis told reporters, "Well I don't know how much of a pattern it is, we actually had seen quite an improvement in Iran's behavior until recently."[43]

Taking on the IRGC made sense to American naval intelligence, which has been aware of the Iranian maritime threat for years. A few unclassified reports, including a Rand Corporation general study of A2/AD, throws light on the possible American responses.[44] Still, the Pentagon has been worried about Iranian efforts to secure A2/AD capabilities in Bab al-Mandeb, one of the choke points that the Triangle Maritime Plan included. Further afield, the MIW calls for mobilizing the network of the Shiite proxy groups and even criminal gangs and mercenaries that have operated off the Somali coast. As will be more fully elaborated in Chap. 7, the Houthi rebellion in Yemen has given NEDSA an opportunity to stage A2/AD exercise in the Red Sea.

deed; IRGC's timely job; Americans with hands on heads, *Khamenei Official Twitter Account*, January 24, 2016, https://twitter.com/khamenei_ir/status/691230624812777473/photo/1.

[43] Reuters, A swarm of Iranian fast-attack boats, forced a U.S. Navy ship to change course in the Persian Gulf, *Reuters*, March 6, 2017, www.businessinsider.com/usns-liberty-iran-force-us-navy-to-change-course-2017-3; Reuters, U.S. Navy ship changes course after Iran vessels come close: U.S. official, March 6, 2017, www.reuters.com/article/us-usa-iran-navy-idUSKBN16D1X3; TIP, "U.S.-Iran Naval Encounters," The Iranian Primer, United States Institute of Peace (TIP), September 7, 2016.

[44] John Gordon IV and John Matsumura, The Army's Role in Overcoming Anti-Access and Area Denial Challenges, RAND Corporation, 2013, www.rand.org/content/dam/rand/pubs/chresearch_reports/RR200/RR229/RAND_RR229.pdf.

Improving Iran's Human Right Record: Another Mission Impossible?

While by no means the worst human rights offender, the record of the Islamic Republic has concerned the international community for decades. Iran has been censured by the United States, the European Union, and the United Nations. In 1985, Iran was "the fourth country ever in the history of the United Nations" to be placed on the agenda of the General Assembly because of the "severity and extent of its abuses of human rights."[45]

Iran has been accused of a wide range of violations: minority rights, gender rights, gay rights, religious rights, civil rights, and political rights. The regime has been known for its mistreatment and torture of prisoners, extrajudicial killings, excessive use of capital punishment, and harsh sentencing guidelines, even for minors. During the tenure of Mahmoud Ahmadinejad, 2005–2013, Iran's human-rights record deteriorated badly. Human Rights Watch reported that executions increased from 86 in 2005 to 317 in 2007. Despite Rouhani's promise to improve this lamentable record, things have hardly changed. Ahmed Shaheed, the UN special rapporteur on human rights in Iran, noted that in 2015 the number of executions reached 977, compared to 743 the year before. Other forms of persecution and harassment have also increased. The statistics for the first half of 2016 was even worse. On July 29, the Iran Human Rights Documentation Center and the Abdorrahman Boroumand Foundation, an independent institution monitoring human rights in Iran, reported 386 executions.[46]

[45] Cooper, R. The Bahá'ís of Iran: The Minority Rights Group Report 51 (London, UK: The Minority Rights Group LTD, 1995).

[46] HRW, UN: Hold Ahmadinejad Accountable for Iran Rights Crisis, *Human Rights Watch*, September 17, 2008, https://www.hrw.org/news/2008/09/17/un-hold-ahmadinejad-accountable-iran-rights-crisis; CHRI, Rights Crisis Escalates Faces and Cases from Ahmadinejad's Crackdown, *Center for Human Rights in Iran*, September 20, 2008, https://www.iranhumanrights.org/2008/09/irancrackdown/; Amnesty International, Death Penalty 2015: Alarming Surge In Recorded Executions Sees Highest Toll in More Than 25 Years, April 6, 2016, https://www.amnesty.org/en/press-releases/2016/04/alarming-surge-in-recorded-executions-sees-highest-toll-in-more-than-25-years/; IHRDC, IHRDC Chart of Executions by the Islamic Republic of Iran – 2016, *Iran Human Rights Documentation Center*, 2016, www.iranhrdc.org/english/publications/human-rights-data/chart-of-executions/1000000620-ihrdc-chart-of-executions-by-the-islamic-republic-of-iran-2016.html.

Long and arbitrary detentions of "peaceful activists, journalists, students and human rights defenders," who had often been charged with "acting against national security" are common. In 2008 the UN General Assembly voted to express "deep concern" for Iran's human rights record particularly "cases of torture, the high incidence of executions and juvenile executions…the persecution of women seeking their human rights, discrimination against minorities and attacks on minority groups like the Bahais in state media." Protests over the "stolen election" of 2009 saw scores killed, hundreds arrested (including dozens of opposition leaders), and a wholesale suppression of the media.[47]

During the re-election campaign, Rouhani focused on the coercive practice of the clerical establishment. He went so far as to mock them, stating: "You can't send people to heaven by the whip." Unfortunately for the moderates, however, the hardline Principalists have virtual control over human rights issues. Following the revolution, the judicial system was restructured to comply with the principles of sharia, although some types of civil status were retained. In addition to the network of public courts, both civil and criminal, the Islamic Revolutionary Courts oversee ill-defined offenses, such as crimes against national security and crimes that undermine the Islamic Republic. The Special Clerical Courts serve to prosecute dissident clerics and others accused of a "theological danger" to the regime. They are independent of the regular judiciary system, accountable to the Supreme Leader alone, and there is no appeal of the sentences.[48]

[47] Ali Akbar Dareini, "Iran Accuses U.S. of Meddling after Disputed Vote," *International Campaign for Human Rights in Iran*, June 17, 2009, http://www.google.com/hostednews/ap/article/ALeqM5jGSJEAPs_r2T2wxsL5G3t4z-jajQD98SIN8O1; Borzou Daragahi, "Hardliners Threaten to Depose Ahmadinejad over Defiance," *Los Angeles Times*, July 29, 2009, latimes.com/2009/jul/29/world/fg-iran29; "UN Secretary General's Report on the Situation of Human Rights in Iran," *International Campaign for Human Rights in Iran*, February 16, 2015, http://www.iranhumanrights.org/themes/documents/un-secretary-general-report.html; "United Nations General Assembly 'Deeply Concerned' about Human Rights Conditions," *International Campaign for Human Rights in Iran*, December 18, 2008, http://www.iranhumanrights.org/themes/news/single-news/article/united-nations-general-assembly-deeply-concerned-about-human-rights-conditions.html; and The National, "Iran Reformists Arrested after Tehran Riots," *The National*, June 14, 2009, www.thenational.ae/news/world/middle-east/iran-reformists-arrested-after-street-protests.

[48] Tabnak, You can't send people to heaven by the whip, *Tabnak.ir*, April 20, 2016, www.tabnak-gilan.ir/fa/news/215602/واکنش-روحانی-به-راهاندازی-گشت-اخلاقی-نامحسوس-وظیفه-ما-در-قبال-مردم-احترام-به-کرامت-آنهاست.

Several Revolutionary Court justices, such as Abolghassem Salavati, Mohammad Moghiseh, Yahya Pirabbassi and Hassan Zareh Dehnavi, have notorious track records in handing down death sentences or long prison terms. Pirabbassi has been known to take an especially activist stand against human rights lawyers, students, journalists, and the Christian and Bahai minorities. Salavati has been assigned cases of dual-nationality Iranians.[49]

The plight of American and Canadian dual nationals generated negative publicity, a fact that led President Obama to arrange for the release of some of them around the implementation of the JCPOA. Since then a new crop of detainees has attracted public attention, including Siamak and Bahger Namazi, Karen Wafadari and his wife Afarin Neisari, and Xiyue Wang, a graduate student at Princeton University.[50]

In yet another peculiarity of the political order, hardline elites have virtually exclusive control in the extensive surveillance and penal system that underpins the Islamic Republic. Ansar e-Hezbollah, a radical group affiliated with the Basij, has reaffirmed its vow to carry out motorbike patrols in Tehran's streets to enforce Islamic norms and "promote virtue and prevent vice." On April 18, 2016, Gasht-e Ershad (Guidance Patrols), supported by the Basij militia, the main agency tasked with enforcing Iran's Islamic code of conduct in public, launched a renewed campaign to force observance of Islamic norms of behavior in public or even private spaces. As before, enforcing the hijab is high on their agenda. The Guidance Patrols have the right to admonish suspects, impose fines, and arrest people. The Basij also is empowered to destroy satellite dishes and receivers as part of a widespread crackdown against devices they say "deviate [sic] morality and culture."[51]

[49] Iradj Mesdaghi, "Judge Salavati One of the Agents of Crime against Humanity," *Iran Global*, August 31, 2013, www.iranglobal.info/node/24337.

[50] Thomas Erdbrink, Tehran Sentences Iranian-American Businessman to 10 Years, Report Says, *The New York Times*, October 18, 2016, https://www.nytimes.com/2016/10/19/world/asia/siamak-namazi-iran-prison.html; CHRI, Grave and Baseless New Charges Against Imprisoned Iranian-American and Wife, *Center for Human Rights in Iran*, March 10, 2017, https://www.iranhumanrights.org/2017/03/new-charges-against-imprisoned-iranian-american/; Rick Gladstone, Iran Sentences U.S. Graduate Student to 10 Years on Spying Charges, *The New York Times*, July 16, 2017, https://www.nytimes.com/2017/07/16/world/middleeast/iran-rouhani-brother-arrest-united-states-nuclear.html.

[51] Kayhan Newspaper, "Mobilizing of 7 Thousand Police for Morality Police," *Kayhan Newspaper*, https://goo.gl/wYq2vT; "Destroying 100,000 Satellite Dishes," *Etemad Newspaper*, July 24, 2016, etemadnewspaper.ir/Default.aspx?News_Id=49,721.

In a highly controversial move, Ansar activists have also blocked women from entering sporting arenas; Haqqani circle theologians argue that letting women into sports stadiums is anti-Islamic and would promote prostitution. Iravani Religious Seminary, which is part of the Haqqani Circle network, teamed up with Ansar-e-Hezbollah members to distribute leaflets around Tehran calling for action to ensure women were kept out of stadiums. Some of the leaflets they distributed warned of a "bloody uprising" if their demands were not accepted. As a rule, attempts to prevent women from enjoying sports have been a barometer of attitudes toward the Sharia-compliant lifestyle.[52]

Not surprisingly, the inability of the Rouhani administration to curb human rights abuses has generated widespread disappointment both domestically and abroad. The liberal United States Institute of Peace, normally sympathetic to Iranian moderates, issued a report entitled "Rouhani Failed on Human Rights Reform." More to the point, the Republican Congress and the Trump administration have promised to apply more pressure on Tehran. Pending legislation is expected to tighten the scrutiny of Iran's human rights record under the Global Magnitsky Act. Originally enacted to punish Russia for the killing of Sergey Magnitsky, a lawyer who exposed corruption among the Russian oligarchs connected to Kremlin, the Act expands the penalties for individuals responsible for human rights violation.[53]

Looking like a "who is who" in the parastatal domain, the State Department's roster of human rights violators reflects the difficulties of the Rouhani government to improve the human rights record. IRGC commanders top the list. Among them are General Mohammad Ali Jafari, and his deputy, Hossein Taeb, Abdullah Araghi Ground Forces Deputy Commander, Mohammad Reza Naqdi, commander of the Basij and his deputy, Ali Fazli. Top intelligence officials such as Heydar Moslehi, the former head of MOIS, and Asghar Mir-Hejazi, the intelligence adviser to the Supreme Leader are also cited. Also, high on the list are judicial officials, including Gholam Hossein Mohseni-Ejei, Judiciary Spokesman and

[52] "Rouhani Reaction to Secret Moral Patrol," *Tabnak*, July 2, 2016, https://goo.gl/gJeXnZ; Ansar e-Hezbollah, "If Women Come to Stadium, We Will Make Bloody Uprising," https://goo.gl/kVvGB6.

[53] Lisa Canak, Rouhani Fails on Human Rights Reforms, *The Iran Primer*, August 12, 2016, iranprimer.usip.org/blog/2016/jun/14/rouhani-fails-human-rights-reforms; Ilan Berman, Trump's arsenal against Iran: Column, *USA Today*, December 29, 2016, https://www.usatoday.com/story/opinion/2016/12/29/iran-trump-magnitsky-nuclear-deal-column/95926188/.

former prosecutor general of Iran, Sohrab Suleimani, supervisor of the Office of the Deputy for Security and Law Enforcement of the State Prisons Organization and former Director General of the Tehran Prisons Organization.[54]

Much as this formidable list posed a problem for relations with the United States, human rights were only a minor concern in Iran's dealings with Russia and the European Union.

Conclusion

Virtually immediately after the revolution, the Islamist regime declared the United States to be the "Great Satan." The Supreme Leader Ayatollah Ruhollah Khomeini decreed that relations with the United States amounted to national treason. Subsequent leaders had tried and failed to balance the ideological dictate, including the Mohammed Khatami who tried to launch the "Dialogue Among Civilizations" outreach. When Hassan Rouhani came to power in 2013, he launched a new outreach bolstered by two advantageous developments. First, the economy was reeling from years of harsh sanctions imposed by the United Nations to punish Iran for proliferation, and second, the election of President Barak Obama, who saw a historic opportunity to heal the poisonous interaction between the two countries. Capped by the JCPOA, the détente promised to open a new chapter in Iran's engagement with the international community as well.

The Principalists, however, bitterly resented the nuclear deal and were even more suspicious of the idea of rejoining the community of nations, a step that the Haqqani circle leader Ayatollah Mohammed-Taghi Mesbah Yazdi considered blasphemous. The Principalists were also able to utilize the profound critique of the JCPOA in the United States, which both Donald Trump and the Republicans turned into a successful campaign line.

With new sanctions looming on the horizon and the failure of Rouhani to improve the economic mismanagement and corruption, the experiment in normalization is arguably over. Although Principalists are not expected to break the JCPOA, they may try the so-called "sneak out," a return to a clandestine effort to weaponize. The naval branch of the Revolutionary Guards,

[54] Farhad Rezaei, Will Iran's Human-Rights Failures Bring Down Its Reformer President? *The National Interest*, December 11, 2016, nationalinterest.org/feature/will-irans-human-rights-failures-bring-down-its-reformer-18685?page=show.

the NEDSA, which has harassed U.S. vessels in the Gulf, could further inflame tensions, especially if the Trump administration reacts with force.

References

Ali Akbar Dareini, "Iran Accuses U.S. of Meddling after Disputed Vote," *International Campaign for Human Rights in Iran*, June 17, 2009, http://www.google.com/hostednews/ap/article/ALeqM5jGSJEAPs_r2T2wxsL5G3t4z-jajQD98SIN8O1.

Amnesty International, Death Penalty 2015: Alarming Surge In Recorded Executions Sees Highest Toll in More Than 25 Years, April 6, 2016, https://www.amnesty.org/en/press-releases/2016/04/alarming-surge-in-recorded-executions-sees-highest-toll-in-more-than-25-years/.

An Overview of the IRGC Speedboats, *Mashregh News*, https://goo.gl/rGiDx9.

Ansar e-Hezbollah, "If Women Come to Stadium, We Will Make Bloody Uprising," https://goo.gl/kVvGB6.

Awarding Fath Medal to the Commander who Arrested American Invaders," *Ayatollah Khamenei's Website*, September 26, 2016, farsi.khamenei.ir/news-content?id=32124.

BBC, "Iran's banks to be blocked from global banking system," *BBC News*, March 15, 2012, http://www.bbc.com/news/business-17390456.

Borzou Daragahi, "Hardliners Threaten to Depose Ahmadinejad over Defiance," *Los Angeles Times*, July 29, 2009, latimes.com/2009/jul/29/world/fg-iran29.

CHRI, Grave and Baseless New Charges Against Imprisoned Iranian-American and Wife, *Center for Human Rights in Iran*, March 10, 2017, https://www.iranhumanrights.org/2017/03/new-charges-against-imprisoned-iranian-american/.

CHRI, Rights Crisis Escalates Faces and Cases from Ahmadinejad's Crackdown, *Center for Human Rights in Iran*, September 20, 2008, https://www.iranhumanrights.org/2008/09/irancrackdown/.

CISADA, "Comprehensive Iran Sanctions, Accountability, and Divestment Act of 2010 (CISADA)," *The Department of the Treasury*, July 1, 2010, http://www.treasury.gov/resource-center/sanctions/Documents/hr2194.pdf.

CNN, Iranian surveillance drone flies over U.S. aircraft carrier in Persian Gulf, January 29, 2016, edition.cnn.com/2016/01/29/politics/iran-drone-uss-harry-truman/?utm_source=Sailthru&utm_medium=email&utm_campaign=New%20Campaign&utm_term=%2ASituation%20Report.

CNN, See Iranian rocket fly by U.S. carrier edition.cnn.com/videos/us/2016/01/09/iran-rocket-us-ship-video-orig-vstan.cnn.

Cooper, R. The Bahá'ís of Iran: The Minority Rights Group Report 51 (London, UK: The Minority Rights Group LTD, 1995).

Council Regulation, "Council Regulation (EU) No 267/2012 of 23 March 2012 Concerning Restrictive Masures Aainst Iran and Repealing Regulation (EU) No 961/2010," *Official Journal of the European Union* 23 March 2012, http://eur-lex.europa.eu/LexUriServ/LexUriServ.do?uri=OJ:L:2012:088:0001:0112:en:PDF.

David Sanger, & Broad, In Iran Talks, U.S. Seeks to Prevent a Covert Weapon, *The New York Times*, November 22, 2014.

EN, "Destroying 100,000 Satellite Dishes," *Etemad Newspaper*, July 24, 2016, etemadnewspaper.ir/Default.aspx?News_Id=49721.

Executive Order 13608, Executive Order 13608: Prohibiting Certain Transactions With and Suspending Entry Into the United States of Foreign Sanctions Evaders With Respect to Iran and Syria, Presidential Documents, Washington, DC: U.S. Government Publishing Office, Vol. 77, 1 May 2012.

Farhad Rezaei, Iran's Ballistic Missiles Program: Changing Course in the Trump Era? Harvard Belfer Center, *Iran Matters*, May 08, 2017, http://www.belfercenter.org/publication/irans-ballistic-missiles-program-changing-course-trump-era.

Farhad Rezaei, Will Iran's Human-Rights Failures Bring Down Its Reformer President? *The National Interest*, December 11, 2016, nationalinterest.org/feature/will-irans-human-rights-failures-bring-down-its-reformer-18685?page=show.

FNA, Negotiation with the West is Betrayal, *Fars News Agency*, October 28, 2016, www.farsnews.com/13950708000252.

Fox News, A Navy destroyer fires warning shots at Iranian vessels in Strait of Hormuz, January 09, 2017, www.foxnews.com/world/2017/01/09/navy-destroyer-reportedly-fires-warning-shots-at-iranian-vessels-in-strait-hormuz.html.

Fox News, Iran threatened to shoot down U.S. Navy spy planes in the Persian Gulf, September 13, 2016, www.foxnews.com/world/2016/09/13/iran-threatened-to-shoot-down-us-navy-spy-planes-in-Persian-gulf.html.

Fox News, Iranian small boat pointed weapon at U.S. Navy Helicopter in Strait of Hormuz, officials say, November 28, 2016, www.foxnews.com/world/2016/11/28/iranian-small-boat-pointed-weapon-at-us-navy-helicopter-in-strait-hormuz-officials-say.html.

Fox News, Navy destroyer has close encounter with Iran vessel in the Persian Gulf, April 25, 2017, www.foxnews.com/world/2017/04/25/navy-destroyer-has-close-encounter-with-iran-vessel-in-persian-gulf.html.

GAO, U.S. and International Sanctions Have Adversely Affected the Iranian Economy: Report to the Chairman, Committee on Homeland Security and Governmental Affairs, U.S. Senate, Washington, DC, GAO-13-326, February 2013.

God's deed: IRGC's timely job: Americans with hands on heads, *Khamenei Official Twitter Account*, January 24, 2016, https://twitter.com/khamenei_ir/status/691230624812777473/photo/1.

H.R. McMaster, Probing for Weakness, *The Wall Street Journal*, March 23, 2016, https://www.wsj.com/articles/probing-for-weakness-1458775212.

Heroux, Is Iran's Nuclear Program Really Getting a 24-Day Pass? *Huffington Post*, August 6, 2015.

Hersh, "The coming wars," *The New Yorker*, 2005, http://www.newyorker.com/magazine/2005/01/24/the-coming-wars.

HRW, UN: Hold Ahmadinejad Accountable for Iran Rights Crisis, *Human Rights Watch*, September 17, 2008, https://www.hrw.org/news/2008/09/17/un-hold-ahmadinejad-accountable-iran-rights-crisis.

IAEA Report by the Director General, Verification and monitoring in the Islamic Republic of Iran in light of United Nations Security Council resolution 2231 (2015), GOV/2017/10 February 24, 2017.

ICHRI, "UN Secretary General's Report on the Situation of Human Rights in Iran," *International Campaign for Human Rights in Iran*, February 16, 2015, http://www.iranhumanrights.org/themes/documents/un-secretary-general-report.html.

ICHRI, "United Nations General Assembly 'Deeply Concerned' about Human Rights Conditions," *International Campaign for Human Rights in Iran*, December 18, 2008, http://www.iranhumanrights.org/themes/news/single-news/article/united-nations-general-assembly-deeply-concerned-about-human-rights-conditions.html.

IHRDC, IHRDC Chart of Executions by the Islamic Republic of Iran – 2016, *Iran Human Rights Documentation Center*, 2016, www.iranhrdc.org/english/publications/human-rights-data/chart-of-executions/1000000620-ihrdc-chart-of-executions-by-the-islamic-republic-of-iran-2016.html.

Ilan Berman, Trump's arsenal against Iran: Column, *USA Today*, December 29, 2016, https://www.usatoday.com/story/opinion/2016/12/29/iran-trump-magnitsky-nuclear-deal-column/95926188/.

Iradj Mesdaghi, "Judge Salavati One of the Agents of Crime against Humanity," *Iran Global*, August 31, 2013, www.iranglobal.info/node/24337.

Iran Newspaper, the Consequences of Iran's Missile Test, *Iran Newspaper*, February 2, 2017, www.iran-newspaper.com/Newspaper/BlockPrint/169767.

Ishaan Tharoor, "The historic nuclear deal with Iran: How it works," *The Washington Post*, July 14, 2015, https://www.washingtonpost.com/news/worldviews/wp/2015/07/14/the-historic-nuclear-deal-with-iran-how-it-works/.

ISNA, America is our Age's Pharaoh, *Iranian Students News Agency*, September 3, 2016, http://www.isna.ir/news/95061308226/.

ISNA, America Seeks to Instill Its Vulgar Culture in Iran, *Iranian Students News Agency*, October 30, 2016, http://www.isna.ir/news/95080906101/.

JCPOA, "Joint Comprehensive Plan of Action," July 14, 2015, https://www.justsecurity.org/wp-content/uploads/2015/07/271545626-Iran-Deal-Text.pdf.

John Gordon IV and John Matsumura, The Army's Role in Overcoming Anti-Access and Area Denial Challenges, RAND Corporation, 2013, www.rand.org/content/dam/rand/pubs/chresearch_reports/RR200/RR229/RAND_RR229.pdf.

Karbaschi Criticizes Iran's Military Presence in the Syrian War, *YouTube*, April 29, 2017, https://www.youtube.com/watch?v=wYW0qr2pHf0.

Karen DeYoung, Trump administration says it's putting Iran 'on notice' following missile test, February 1, 2017, https://www.washingtonpost.com/world/national-security/2017/02/01/fc5ce3d2-e8b0-11e6-80c2-30e57e57e05d_story.html?utm_term=.b1a66e19d1a2.

Kayhan Newspaper, "Mobilizing of 7 Thousand Police for Morality Police," *Kayhan Newspaper*, https://goo.gl/wYq2vT.

Kayhan Newspaper, Ayatollah Khamenei Called America "Great Satan," *Kayhan Newspaper*, June 6, 2015, kayhan.ir/fa/news/46577.

Kayhan Newspaper, Historical Turn and Uts Requirements, *Kayhan Newspaper*, February 1, 2017, kayhan.ir/fa/mobile/news/96856/927.

Kenneth Katzman, Iran sanctions (Washington: Congressional Research Service, 2011).

Lisa Canak, Rouhani Fails on Human Rights Reforms, *The Iran Primer*, August 12, 2016, iranprimer.usip.org/blog/2016/jun/14/rouhani-fails-human-rights-reforms.

Moshe Ya'alon, Current Iran framework will make war more likely, *The Washington Post*, April 8, 2015.

PBS, "Stuck Between America and Iran," *PBS, Tehran Bureau*, September 18, 2009, http://www.pbs.org/wgbh/pages/frontline/tehranbureau/2009/09/stuck-between-america-and-iran.html.

Reuters, A swarm of Iranian fast-attack boats, forced a U.S. Navy ship to change course in the Persian Gulf, *Reuters*, March 6, 2017, www.businessinsider.com/usns-liberty-iran-force-us-navy-to-change-course-2017-3.

Reuters, After nuclear deal, Iran Sticks to Wary Tactics in Key Gulf Waterway, July 12, 2016, www.reuters.com/article/us-usa-iran-warship-idUSKCN0ZR2E0.

Reuters, Iran's vessels make 'high-speed intercept' of U.S. ship: U.S. official August 25, 2016, www.reuters.com/article/us-usa-iran-military-idUSKCN10Z2OP.

Reuters, Trump election puts Iran nuclear deal on shaky ground, *Reuters*, November 9, 2016, www.reuters.com/article/us-usa-election-trump-iran-idUSKBN13427E.

Reuters, U.S. Navy ship changes course after Iran vessels come close: U.S. official, March 6, 2017, www.reuters.com/article/us-usa-iran-navy-idUSKBN16D1X3.

Reuters, U.S. Navy ship targeted in failed missile attack from Yemen: U.S., October 10, 2016, www.reuters.com/article/us-yemen-security-usa-ship-id USKCN12A082?il=0.

Rex W. Tillerson, Trump Administration Undergoing Interagency Review of Iran Deal, April 18, 2017, https://www.state.gov/secretary/remarks/2017/04/270315.htm.

Rick Gladstone and Stephen Castle, Global Network Expels as Many as 30 of Iran's Banks in Move to Isolate Its Economy, *The New York Times*, March 15, 2012.

Rick Gladstone and Stephen Castle, Global Network Expels as Many as 30 of Iran's Banks in Move to Isolate Its Economy, Brendan Taylor, "Chapter Three: Sanctioning Iran," *The Adelphi Papers*, Vol. 49, No. 411 (2009).

Rick Gladstone, Iran Sentences U.S. Graduate Student to 10 Years on Spying Charges, *The New York Times*, July 16, 2017, https://www.nytimes.com/2017/07/16/world/middleeast/iran-rouhani-brother-arrest-united-states-nuclear.html.

Rooz Online, Democracy and Human Rights have no Place in Islam, *Rooz Online*, September 1, 2010, www.roozonline.com/persian/news/newsitem/article/br-9.html.

Ruhollah Khomeini: In this Revolution, America is the Great Satan, Sahife Imam, Vol. 10.

Tabnak, Rouhani Reaction to Secret Moral Patrol, *Tabnak*, July 2, 2016, https://goo.gl/gJeXnZ.

Tabnak, You can't send people to heaven by the whip, *Tabnak.ir*, |April 20, 2016, www.tabnakgilan.ir/fa/news/215602/ واکنش-روحانی-به-راه-اندازی-گشت-اخلاقی-نامحسوسوظیفه-ما-
در-قبال-مردم-احترام-به-کرامت-آنهاست.

TCACNP, "Factsheet: Iran and the Additional Protocol," *The Center for Arms Control and Non-proliferation*, July 14, 2015, armscontrolcenter.org/factsheet-iran-and-the-additional-protocol.

The National, "Iran Reformists Arrested after Tehran Riots," *The National*, June 14, 2009, www.thenational.ae/news/world/middle-east/iran-reformists-arrested-a fter-street-protests.

Thomas Erdbrink, Iran Staggers as Sanctions Hit Economy, *The New York Times*, September 30, 2013.

Thomas Erdbrink, Tehran Sentences Iranian-American Businessman to 10 Years, Report Says, *The New York Times*, October 18, 2016, https://www.nytimes.com/2016/10/19/world/asia/siamak-namazi-iran-prison.html.

TNA, Defense Minister, The Recent Missile Test was Successful, February 1, 2017, *Tasnim News Agency*, https://www.tasnimnews.com/fa/news/1395/11/14/1315991/وزیر-دفاع-آزمایش-موشکی-اخیر-موفقیت-آمیز-بود.

TNA, Iran's Supreme Leader Emphasized that Tehran will Negotiate on Nuclear Issue Only, *Tasnim News Agency*, March 21, 2015, https://www.tasnimnews.com/fa/news/1394/01/01/692344/رهبر-ایران-تأکید-کرد-که-تهران-
تنها-درباره-موضوع-هسته-ای-با-آمریکا-مذاکره-می-کند

Tower, "Deal Will Give Iran 24 Days to Cover Up Suspected Nuclear Sites," *TheTower.org*, July 14, 2015, http://www.thetower.org/2296-deal-will-give-iran-24-days-to-cover-up-suspected-nuclear-sites/.

TWP, Tillerson's full statement on Iran, *The Washington Post*, April 19, 2017, https://www.washingtonpost.com/video/national/watch-tillersons-full-statement-on-iran/2017/04/19/4a1bc8ba-2540-11e7-928e-3624539060e8_video.html.

U.S. Sailor Flap: Model for Diplomacy? January 12, 2016, iranprimer.usip.org/blog/2016/jan/12/10-us-sailors-held-iran-be-returned.

UNSC Resolution 1737, Adopted by the Security Council at its 5612th meeting on the Non-Proliferation of Nuclear Weapons, Official Record, S/RES/1737, New York: United Nations Security Council (UNSC), 23 December 2006.

UNSC Resolution 2231 (2015), Adopted by the Security Council at its 7488th meeting, on 20 July 2015, United Nations Security Council, S/RES/2231 (2015).

UNSC Report, "Chronology of Events: Iran," *Security Council Report.org*, December 1, 2014, http://www.securitycouncilreport.org/chronology/iran.php?page=1, Executive Order 13599, Executive Order 13599 of February 5, 2012: Blocking Property of the Government of Iran and Iranian Financial Institutions, Washington, DC: Presidential Documents, Vol. 77, 8 February 2012.

UNSC Resolution 1803, Adopted by the Security Council at its 5848th meeting on the Non-Proliferation of Nuclear Weapons, Official Record, S/RES/1803, New York: United Nations Security Council (UNSC), 3 March 2008.

UNSC Resolution 1929, Adopted by the Security Council at its 6335th meeting on the Non-Proliferation of Nuclear Weapons, Official Record, S/RES/1929, New York: United Nations Security Council (UNSC), 9 June 2010.

USDS, Strategic Assessment Bureau of Counterterrorism and Countering Violent Extremism, Country Reports on Terrorism 2015, Chap. 1, www.state.gov/j/ct/rls/crt/2015/257513.htm.

USNI, Iranian Boats Harass Another U.S. Navy Patrol Coastal Ship In Persian Gulf, September 6, 2016, https://news.usni.org/2016/09/06/iranian-boats-harass-another-u-s-navy-patrol-coastal-ship-persian-gulf.

USNI, Three More U.S. Navy Ships Harassed by Iranian Patrol Boats, August 25, 2016, https://news.usni.org/2016/08/25/three-u-s-navy-ships-harassed-iranians-patrol-boats.

Yvette Hovsepian-Bearce, The Political Ideology of Ayatollah Khamenei. Out of the Mouth of the Supreme Leader (London: Routledge, 2016).

CHAPTER 3

Iran and Russia: Completing the Pivot to the East?

Unlike the United States, which the Khomenist ideologues considered an implacable enemy of Islam, relations with Russia and its predecessor the Soviet Union were more nuanced. During the Cold War, Iran's policy toward its neighbors to the East was significantly influenced by its alliance with the Western bloc led by the United States. Ayatollah Ruhollah Khomeini denounced both Washington and Moscow in his famous "neither East nor West" address. As he put it, "It is not secret that the durability and consistency of the Islamic Republic of Iran is based on the policy of neither the East nor the West, and the abandonment of this policy will be betrayal of Islam and Muslims and the decline of the dignity, credibility, and independence of the country and the brave nation of Iran." But other leaders were more attuned to neorealist considerations, as noted in Chap. 1.[1]

Akbar Hashemi Rafsanjani, who dictated the foreign policy of the new regime, was keen to pursue economic relations with the Soviet Union. In his views, Iran, faced with a ruinous war with Iraq estimated at some $8 billion annually, had little choice but to turn to Moscow. Rafsanjani's ambitious plan to restart construction of the Shah-era Bushehr reactor was a key driver in the decision. The Soviet leadership in Moscow, looking for new sources of revenue, was highly receptive. The Soviet deputy prime minister

[1] Rouhollah.ir, The General Policies of the Islamic System During the Rebuilding of the Country, October 2, 1988, farsi.rouhollah.ir/library/sahifeh?volume=21&page=155.

F. Rezaei, *Iran's Foreign Policy After the Nuclear Agreement*, Middle East Today, https://doi.org/10.1007/978-3-319-76789-5_3

Georgi Kornyenko visited Tehran on February 26, 1986, and agreed to "expand economic and trade relations, and to conduct joint oil exploration in the Caspian Sea." Rafsanjani lauded Kornyenko's visit, stating that it "will have a great effect on our relations with the Soviet Union and the Eastern world... One can be optimistic in fields such as technical, military, economic and possibly political relations."[2]

Economics aside, the relations were politically beneficial. The regime pragmatists hoped to secure Moscow's assistance in the war with Iraq, or at least dissuade it from a single support of Saddam Hussein, the Iraqi dictator. The Soviet leaders wished to placate Ayatollah Khomeini who spoke out very harshly against their involvement in Afghanistan. But the Supreme Leader, incensed about the Soviet invasion of a Muslim country, was not easily dissuaded from tirades against the "atheist East" that, in his view, was as evil as the "infidel West." Publicly revealing the extent of burgeoning economic and political ties between the two countries had to wait until the death of Khomeini on June 3, 1989. Less than three weeks later, on June 20, Rafsanjani traveled to Moscow where Mikhail Gorbachev prepared an impressive official visit. The Soviet leader called the visit a "landmark event" in international relations and promised that "there cannot and will not be anything in our policy that would damage Iran's interests." For his part, Rafsanjani, who became president on August 3, 1989, pledged to make the turn to the East a permanent feature of Iran's foreign policy.[3]

Hiding behind Rafsanjani rhetoric was a plan to harness the Soviet nuclear know-how to expand Iran's nuclear project, which had stalled because European countries, under pressure from the United States, had refused to help. On June 22, 1989, Rafsanjani and Gorbachev signed an agreement on nuclear cooperation, followed by a March 1990 comprehensive deal to complete the Bushehr project, among others.[4]

[2] Martin Sicker, The Bear and the Lion: Soviet Imperialism and Iran (New York: Praeger Publishers, 1988), p. 128; Reuters, Around The World; Iranian Says Ties With Soviet Improve, *The New York Times*, February 10, 1986, www.nytimes.com/1986/02/10/world/around-the-world-iranian-says-ties-with-soviet-improve.html; Cameron R. Hume, The United Nations, Iran, and Iraq: How Peacemaking Changed (Indiana University Press, 1994).

[3] Michael Parks, "Iranian Gets Warm Soviet Welcome: Rafsanjani-Gorbachev Talks Point to Improved Relations," *Los Angeles Times*, June 21, 1981.

[4] Hibbs, "German officials say governments may not let Bushehr be finished," *Nucleonics Week*, Vol. 32, No. 6 (1991), p. 15; Tyler, Rebuilding Plan to Cost $15 Billion, Tehran Says, *The Washington Post*, July 3, 1989.

In addition to selling civilian nuclear technology, Russia helped Iran modernize its conventional forces and allegedly provided Iran with the know-how and technology for its ballistic missile, chemical, and biological warfare programs. The cooperation was a win-win strategy; while the sale of arms and nuclear reactors to Iran helped Russia's cash-strapped military-industrial complex, Iran got the much-needed modern technology that no other country would supply. The exchange fit the Jahad-e Khodkefayee, the self-reliance strategy that Ayatollah Khomeini had decreed. Moreover, the partnership helped Moscow's ambition in playing a greater role in the Middle East, while Tehran's friendship with Moscow helped the pivot to the East.[5]

Building an Economic and Strategic Alliance with Vladimir Putin

Unlike his predecessor, Boris Yeltsin, who tried to nudge Russia into a pro-Western direction, Vladimir Putin had a different vision for the country upon taking office in 2000. Declaring the world to be multipolar, the Russian leader urged a better relationship with the East, including China, as a counterbalance to the West. For Iran, struggling to break out of its political isolation, the Putin vision came at a fortuitous time. President Mohammed Khatami, who spent a lot of energy on his Western outreach, had been gradually marginalized, as described in Chap. 1, giving the hard-line traditionalists an opportunity to continue with their pivot to the East.

Since coming to power in 2005, Ahmadinejad made the pivot to the East the main pillar of his foreign policy, not least to compensate for the Western sanctions. Geopolitically, the policy included several Asian countries in addition to Russia. Iran signed a bilateral agreement with India, China, and Pakistan,[6] and joined collective organizations such as the Shanghai Cooperation Organization (SCO), Association of Southeast Asia Nation (ASEAN), South Asian Association for Regional Cooperation (SARC), Economic Cooperation Organization (ECO), and D-8 Organization for Economic Cooperation.[7]

[5] Ali Jalali, "The Strategic Partnership of Russia and Iran," *Parameters*, Winter 2001–2002, pp. 98–111.

[6] Mottaki, M, "Iran's Foreign Policy under President Ahmadinejad," *Discourse: An Iranian Quarterly* (Winter 2009).

[7] Farhad Rezaei, Iran's Nuclear Program: A Study in Proliferation and Rollback (New York: Palgrave Macmillan, 2016).

Pivoting to the East made sense to the traditional Principalists led by Ahmadinejad. The 20-Year Perspective Document for Iran spelled out their vision. After completing four consecutive five-year development plans ending in 2025, Iran would be transformed into a fully developed country. This transformation would place her at the top of the 28 nations in the Middle East and Southeast Asia. As a nation with an Islamic and revolutionary identity, Iran would offer a guiding light for the Islamic world and beyond. In the meanwhile, Ahmadinejad hoped to evade the high cost of the nuclear sanctions by linking up with the economies of the East.[8]

Frustration from the West was a part of the pivot to the East. As Ali Akbar Velayati, the former Foreign Minister and an advisor to Ayatollah Ali Khamenei, noted, "there are fundamental obstacles between us, the United States and Europe, which makes it difficult and sometimes impossible to achieve cooperation and coalitions." He added that "cooperation with the West" is not possible except through the granting of many concessions and political retreats. In other words, doing business with the East did not require giving up on revolutionary exports or improving the human rights record, which the West had insisted. Indeed, Velayati implied that both Russia and China "conquered many world markets" without changing their domestic policies, a feat which Iran could reproduce as well.[9]

Alireza Sheikhattar, a close ally of Ahmadinejad, further explained that, unlike Western countries, Eastern countries are not interested in getting involved in [Iran's domestic] political issues. Brigadier General Amir Ali Hajizadeh, Commander of the Revolutionary Guards Aerospace Force, noted, "If we look at the West, we need to take a begging bowl, and link the country's destiny to the Democrats and Republicans. We must constantly retreat, we must constantly see our honors are gone one by one."[10]

[8] Khamenei.ir, the 20-Year Perspective Document for Iran, farsi.khamenei.ir/special?id=4133; Iranian T.V Channel-2, "Iranian President Mahmoud Ahmadinejad: So What If Two U.N. Resolutions Were Issued against Us? America Cannot Cause Iran Real Damage," quoted by Middle East Media Research Institute MEMRI, January 23, 2007, http://www.memritv.org/clip_transcript/en/1369.htm; MacAskill, & Tait, "Europe proposes limited sanctions to halt Tehran's nuclear ambitions," *The Guardian*, April 11, 2006, http://www.theguardian.com/world/2006/apr/11/nuclear.eu; Seyed Hossein Mousavian, The Iranian Nuclear Crisis: A Memoir (Washington: Carnegie Endowment for International Peace, 2012).

[9] Ali Akbar Velayati, Eastern Partnership: The status of the East and West in the foreign policy of the Islamic Republic of Iran, *Hamshahri Diplomatic*, Vol. 1, No. 3.

[10] Alireza Sheikhattar, Accompanying Diplomacy and Economics: The Place of Economic Diplomacy in Regional Cooperation of the Islamic Republic of Iran, *Hamshahri Diplomatic*,

For the Revolutionary Guards and other traditionalists, Russia was an ideal partner for the self-reliance (*Jahad-e Khodefayee*) project, an effort to achieve a degree of independence in the manufacturing and military sectors. The Guards, still scarred by the memory of the arms embargo during the war with Iraq, were especially keen to increase self-reliance in the military industry. Russia also helped Iran to enhance its domestic arms production through the building up of its military-industrial base. Based on a 1991 contract, Iran was able to develop a production line of tanks, military vehicle, and airplanes. According to accounts, Iran produced 1000 × T-72S tanks and 1500 × BMP-2 infantry fighting vehicles, 18 × MiG-29A and 7 × twin-seater MiG-29UB aircraft. Iran attempted to acquire Russian S-300 MPU and Tor-1 air defense and anti-missile systems and a wide range of other mobile surface-to-air missiles (SAMs). Other items that Russia sold Iran as part of the 1991 contract were 25 × Mi-17 medium-lift helicopters, 1000 × Igla-1E shoulder-launched air defense missiles, several battalions of SA-10 and SA-12 surface-to-air missiles, and air-surveillance radars. Russia also sold 3 × Kilo-class attack submarines (877EKM diesel-electric submarines), which could be used to disrupt shipping in the Gulf. The deal also included naval mines, torpedoes, the latest versions of T-90 and T-72 tanks and the Tor-1 SAMs, as well as the Russian Su-27 interceptor fighter for upgrading its air defense cover and Su-25 ground-attack aircraft.[11]

With sales exceeding $4 billion between 1992 and 2000, Iran became the third largest customer for Russian weapons. Further sales included 24 × MiG-29 fighter-bombers; 12 × Su-24MK fighter-bombers, 2 × S-200 air defense systems; 444 × R-27R and 575 × R-60 anti-aircraft missiles for the MiG-29s; 123 × Infantry Fighting Vehicles; 126 × T-72 M1 tanks; 2 × 140 mm artillery systems; 2 × rocket launcher systems and 3 × Kilo-class attack submarines (877EKM diesel-electric submarines), which could be used to disrupt shipping in the Gulf. Moscow has facilitated the sale of technology to Iran that is used in the manufacture of the Soviet-era SS-4 intermediate range ballistic missiles (IRBMs) and has helped Iran to develop its Shahab-3 IRBM, which has a range of 1200 km.[12]

Vol. 1, No. 3. 2006; FNA, If we look at the West, we need to take a begging bowl, *Fars News Agency*, www.farsnews.com/13960207000849.

[11] Ali Jalali, "The Strategic Partnership of Russia and Iran," *Parameters*, Winter 2001–2002, pp. 98–111.

[12] Iran Repairs Five of the Former Iraqi MIG-29 (ИРАН ОТРЕМОНТИРОВАЛ ПЯТЬ БЫВШИХ ИРАКСКИХ МИГ-29), April 16, 2012, http://www.militaryparitet.com/perevodnie/data/ic_perevodnie/2741/%20https://topwar.ru/23840-dizel-elektricheskie-podvodnye-

The Russians have also been assisting Iran's missile programs. The Shahab-3 and Shahab-4 missile projects benefited largely from technology transfers from Russia in addition to the assistance from North Korea. Russian entities also produced critical engine components for the Shahab-3 and Shahab-4 that Iran was unable to produce on its own.[13]

Iran fitted well into Putin's long-term plan to regain Russia's influence in world affairs, starting with the Middle East. Constructing an overland transit corridor has been a key component of this strategy. Development moved ahead mostly below the international radar as Iran extended its domestic north–south rail line from Bandar Abbas, a major Iranian port on the Persian Gulf, north to Qazvin. As part of the North–South Transit Corridor (NSTC), it will soon link Iranian and Russian rail systems through Azerbaijan, the only state bordering both nations. The land route will rival the maritime trade route through the Suez Canal with economic, political, and military repercussions around the globe.[14]

Ayatollah Khamenei, whose security adviser Ali Larijani developed a similar initiative called "looking to the East," went even further. During the visit of Russian Security Council Secretary Igor Ivanov in January 2007, the Supreme Leader proposed that Iran and Russia form an alliance to share responsibilities for the Middle East and Central Asia. When Khamenei visited Moscow in the same month, he urged Moscow to create a natural gas producers' organization like OPEC. Iran argued that creating such an organization would provide a ground for strengthening the gas market, which would lead to coordination between the gas producers and exporter countries, in addition to exchanging technical expertise

lodki-proekta-877-paltus.html; W&P, Russia will consider and 'work out' possible applications of Iran (Россия рассмотрит и "будет прорабатывать" возможные заявки Ирана), *War and Peace*, December 25, 2007, www.warandpeace.ru/ru/news/view/18412/; Diesel-electric submarines of Project 877 "Paltus" (Дизель-электрические подводные лодки проекта 877 «Палтус»), *Boehhoe*, February 5, 2013, https://topwar.ru/23840-dizel-elektricheskie-podvodnye-lodki-proekta-877-paltus.html; Arthur Blinov, Iran in the crossroads of geopolitics (Иран в перекрестье геополитики), www.ng.ru/courier/2006-01-16/14_iran.html?id_user=Y; Ariel Cohen, Countering Russian-Iranian Military Cooperation, *The Heritage Foundation*, April 5, 2001, www.heritage.org/europe/report/countering-russian-iranian-military-cooperation.

[13] Michael Eisenstadt, Russian Arms and Technology Transfers to Iran: Policy Challenges for the United States, *Arms Control Today*, March 2001, https://www.armscontrol.org/print/807.

[14] Audrey L. Altstadt, Kennan Cable No. 23: Putin's Middle East Triangle, *Wilson Center*, June 19, 2017, https://www.wilsoncenter.org/publication/kennan-cable-no23-putins-middle-east-triangle.

between the member states. Khamenei's foreign policy adviser Velayati, who traveled to Moscow in February 2007, expanded on the proposal to ally "against common adversaries and proposed that the two countries share between them the responsibility for the future of the Middle East and Central Asia."[15]

Putin and his foreign policy team were not entirely keen on a formal alliance with Iran, but the Russian president visited Tehran in November 2007 to attend a Caspian Sea summit there. Even so, the Iranian media hailed the visit as a historical milestone and pointed out that both countries "prioritized bilateral cooperation, particularly vis-à-vis the US." As a part of a new energy deal, Iran agreed to let the gas industry company Gazprom, which at the time participated in the development of the second and third phase of the South Field, to engage in subsequent explorations. The Russians also wanted Gazprom to take part in the construction of the Iran–Pakistan–India pipeline, a huge project with political implications in South East Asia; Iranians promised that after the completion of the Bushehr plant, Russia would build two other power plants totaling more than $4 billion.[16]

Russia also agreed to export 50 Turbocharger 53-DR engines for installation in Iran's "Azarakhsh" Iranian supersonic bomber and fighter jets. Also, Russia agreed to sell Iran the 5000-RD engines for its Shafagh aircraft, the latest product of the Iranian airplane industry, which, according to Iranians will be made using the stealth technology. This deal was part of the comprehensive Russian–Iranian agreement, which included 130 economic projects worth more than $100 billion.[17]

President Rouhani's ambitious plans to join the West did not sacrifice the pivot to the East. Like his predecessor, Rouhani considered Iran's

[15] Mark N. Katz, Russian-Iranian Relations in the Ahmadinejad Era, *Middle East Journal*, Vol. 62, No. 2, Spring 2008; Donyaye Eghtesad, The First Step for Creation of Gas OPEC, *Donyaye Eghtesad*, February 3, 2007, donya-e-eqtesad.com/news/665687; NIOC, OPEC Gas, from an understanding to an official organization, *National Iranian Oil Company*, www.nioc.ir/Portal/Home/ShowPage.aspx?Object=NEWS&ID=9d4809fc-88bd-45ea-96eb-74ef0c2e0ae0&LayoutID=d2b4e819-225e-417a-9564-dd14973277fc&CategoryID=5a83de0d-4c67-4b8a-a794-08e669faf33d; The formation of an organization of gas exporting countries, a guarantee for the future of energy trade, *Shana.ir*, April 9, 2007, www.shana.ir/fa/newsagency/101544/تشکیل-سازمان-کشورهای-صادر-کننده-گاز-تضمینی-برای-آینده-تجارت-انرژی.

[16] Mark N. Katz, Russian–Iranian Relations in the Ahmadinejad Era," *Middle East Journal*, Vol. 62, No. 2, Spring 2008.

[17] Why Did Putin Travel to Iran? *Resalat Newspaper*, December 3, 2007, vista.ir/article/286484/چرا-پوتین-به-ایران-سفر-کرد؟.

associate membership in the Shanghai Cooperation Organisation (SCO), a military-economic pact comprising China, Russia, Kazakhstan, Kyrgyzstan, Tajikistan, and Uzbekistan. In August 2014, Iran signed a $20 billion oil deal with Russia to help organize Tehran's oil sales as well as "cooperate in the oil-gas industry, construction of power plants, grids, the supply of machinery, consumer goods and agriculture products."[18]

In 2013–2015, Rouhani and Vladimir Putin met several times, either in formal meetings or casual contacts on the sidelines of another conference. The list was impressive. The two leaders talked at the Council of the SCO in Bishkek on September 13, 2013, and at the summit of the Shanghai Conference on Interaction and Confidence Building Measures in Asia on May 21, 2014. They also held discussions at the Council of SCO heads of state in Dushanbe on September 12, 2014, and at the 4th Caspian Summit in the southern Russian city of Astrakhan on September 29, 2014. On July 9, 2015, Putin and Rouhani met on the sidelines of the BRICS Summit—a conference attended by the heads of state of Brazil, Russia, India, China and South Africa—in the Russian city of Ufa. The leaders discussed key issues affecting Russian–Iranian relations, including prospects for expanded trade and economic ties and energy cooperation. On November 23, 2015, Putin visited Tehran for a summit of the Gas Exporting Countries Forum, where he met Rouhani and other regime leaders.[19]

As the economic cooperation grew, more meetings followed. On August 8, 2016, Rouhani joined Putin and their Azerbaijani counterpart Ilham Aliyev for a trilateral summit in Baku. On March 27, 2017, Rouhani paid a two-day visit to Russia, where Russian Prime Minister Dmitry Medvedev and President Putin signed a string of bilateral agreements to boost trade between the countries more than 70 percent. Testifying to the significance of the relations, Rouhani was received with great pomp and given an hon-

[18] FNA, Vladimir Putin Soon to Sign Historic $20bln Oil Deal with Iran to Bypass Western Sanctions, *Fars News Agency*, August 6, 2014, en.farsnews.com/newstext.aspx?nn=13930515000328; Andrew Tortman, Vladimir Putin signs historic $20bn oil deal with Iran to bypass Western sanctions, *The Telegraph*, August 6, 2014, www.telegraph.co.uk/finance/newsbysector/energy/oilandgas/11014604/Vladimir-Putin-signs-historic-20bn-oil-deal-with-Iran-to-bypass-Western-sanctions.html.

[19] 'A Lot of Work to Do': What's on the Agenda of the Putin-Rouhani Meeting? https://sputniknews.com/world/201703271052004318-russia-iran-putin-rouhani-visit/; Presidents Of Azerbaijan, Iran, And Russia Meet In Baku, https://www.rferl.org/a/azerbaijan-iran-russia-presidents-meet-baku/27909041.html; Russian President Putin arrives in Tehran, www.presstv.ir/Detail/2015/11/23/438822/Russia-Putin-Tehran.

orary doctorate by Moscow State University. In his acceptance speech, the Iranian president emphasized that Moscow and Tehran had "begun effective cooperation" in countering "terrorists" in Syria, in a speech at Moscow State University, which bestowed an honorary doctorate on him. According to the Russian Federal Customs Service, the bilateral trade turnover reached $2.184 billion in late 2016, with Russian exports and imports totaling $1.881 billion and $302.6 million, respectively.[20]

These high-level contacts had important consequences on the ground. In September 2015, the two countries signed a new agreement to deliver two units of S-300PMU-2 air defense systems to Iran worth to over $1 billion. In December 2015, Russia signed a contract to build two more reactors at the Bushehr complex. At the same time, Russia offered a state credit of $5 billion to help Tehran's struggling banking sector. Khamenei praised Putin for "neutralizing Washington's plots" and vowed to expand the economic relations "beyond the current level." The Russia–Iran bilateral trade increased from $1.2 billion in 2015 to almost $2 billion in 2016, a 60 percent growth, which could be termed a breakthrough for Iran.[21]

As before, military sales were an important part of the economic agenda. On February 3, 2016, the Russian company Uralvagonzavod offered to license production of third-generation T-90S—the export version of the T-90A—one of the most advanced main battle tanks in service with the Russian military and the latest development in the T-series of Russian

[20] 'A Lot of Work to Do': What's on the Agenda of the Putin-Rouhani Meeting? https://sputniknews.com/world/201703271052004318-russia-iran-putin-rouhani-visit/; PressTV, Iran president to pay visit to Russia for key talks, *PressTV*, www.presstv.ir/Detail/2017/03/24/515463/Iran-Russia-Hassan-Rouhani-Vladimir-Putin-Parviz-Esmaeili-Moscow; Russia Iran relations and Pres. Rouhani's visit to Moscow, *Tehran Times*, April 8, 2017, https://www.pressreader.com/iran/tehran-times/20170408/281801398819035; *Iran Newspaper*, Strategy of 'Looking to the East' in Rouhani Government, www.iran-newspaper.com/newspaper/BlockPrint/157208.

[21] Russia proposes to Iran to organize licensed production of Russian T-90S tanks, February 03, 2016, tass.com/defense/854186; Domestic Economy, Tehran, Moscow to Sign Investment Guarantee Deal, November 16, 2015, https://financialtribune.com/articles/domestic-economy/30319/tehran-moscow-to-sign-investment-guarantee-deal; Sputnik, Vladimir Putin in the Land of Ayatollahs, *Sputnik*, November 27, 2015, https://sputniknews.com/radio_red_line/201511271030871131-vladimir-putin-in-the-land-of-ayatollahs/; Nikita Smagin, How Russia managed to double its exports to Iran in 2016, *Russia Beyond*, February 17, 2017, https://www.rbth.com/business/2017/02/17/russia-exports-iran-704108.

tanks.[22] On February 16, 2016, Iran's Defense Minister, General Hossein Dehqhan traveled to Moscow to buy more weapons. The list included Sukhoi Su-30SM multirole fighters, the Yakovlev Yak-130 advanced jet trainer and light attack aircraft, and the Mil Mi-8 and Mil Mi-17 transport helicopters, as well as the K-300P Bastion-P mobile coastal defense missile systems, frigates, and diesel-electric submarines.[23] On the same day, Dehqhan announced that Iran and Russia are about to sign a contract for the co-production of an undisclosed number of Russian-made Sukhoi Su-30 multirole fighter aircraft.[24] In April 2016, Russia provided the sophisticated S-300 defense system, to Tehran, a deal due since 2010. Reportedly, Iran devoted a budget of $40 billion for modernizing its army and, hence, Russia could find this market lucrative. In 2016 Russia and Iran signed Memoranda of Understanding (MOU) on dozens of industrial projects, the total amount of $40 billion.[25] This strengthening of relations has come about due to the fact that that both countries were under sanctions because of Russia's foray into the Ukraine. In 2014, just after the Joint Plan of Action (JPA) signed in 2013, Moscow agreed to build eight nuclear power reactors in Iran.[26]

Plans to boost trade relations between Iran and Russia saw an increase of 70 percent, bringing the total to $8.5 billion by May 2016. In the same month, the two countries organized a business forum to discuss prospects for future trade. In July 2016, the Russian energy minister Alexander Novak

[22] Russia proposes to Iran to organize licensed production of Russian T-90S tanks, February 03, 2016, tass.com/defense/854186.

[23] Sputnik, S-300 Was Just the Beginning: Iran Plans Huge Arms Deals With Russia, https://sputniknews.com/military/201602171034893494-russia-iran-military-deals/; Su-30s, Mi-8 s: Iran Considers Inking $8 Billion Arms Deal With Russia, https://sputniknews.com/military/201602161034839542-iran-russia-military-cooperation/; TAAS, Iran's demand for Russian weaponry is estimated at $10 bln, *TASS Russian News Agency*, November 14, 2016, tass.com/defense/912141.

[24] TAAS, Iran's demand for Russian weaponry is estimated at $10 bln, *TASS Russian News Agency*, November 14, 2016, tass.com/defense/912141.

[25] Kommersant, Russia is localized in Iran, [Россия локализуется в Иране], *Kommersant*, https://www.kommersant.ru/doc/3170040; Gevorg Mirzayan, We have agreements for 40 billion, [У нас соглашения на 40 миллиардов], *Expert Online*, expert.ru/2016/07/17/u-nas-soglasheniya-na-40-milliardov/.

[26] Putin encourages Iran to join Russia-led Eurasian alliance, https://www.rt.com/politics/354743-putin-voices-support-to-iran/; Iran will give Russia priority in any industry it wants to invest in, April 21, 2016, https://in.rbth.com/news/2016/04/21/iran-will-give-russia-priority-in-any-industry-it-wants-to-invest-in_586933; Debalina Ghoshal, Russia-Iran: Why is the S-300 Sale Significant? *Institute of Peace and Conflict Studies*, June 19, 2015, www.ipcs.org/article/iran/russia-iran-why-is-the-s-300-sale-significant-4891.html.

promised to invest in Iran's oil and gas projects. Russian truck manufacturer Kamaz signed an agreement with Iran to export 300 trucks in 2017, and GAZ, another Russian manufacturer, signed a memorandum with the Iranian authorities for the supply of 900 buses. In addition, Uralvagonzavod, the main manufacturer of railroad carriages in Russia, won the contract for the supply of 5000 wagons, the first batch of which was shipped in September 2016. On August 22, Dehghan declared that Iran had started negotiations with Russia over the purchase of an unknown number of Sukhoi Su-30SM multi-role fighter jets. In November 2016, Iran and Russia signed a $10 billion arms deal, which included the supply of advanced tanks, artillery systems, planes, and helicopters. In December 2016, Moscow accepted Tehran's conditions for a possible fighter jet deal.[27]

On July 31, 2017, Russia and Iran signed a $2.5 billion deal to start up a rail wagon production operation. The agreement was forged between the Industrial Development and Renovation Organization of Iran (IDRO) and Transmashholding, the Russia's largest rail equipment supplier which will be 80 percent owned by the Russian partner.[28]

[27] Aljazeera, Russia to start building Iran nuclear reactors, www.aljazeera.com/news/2015/12/russia-start-building-iran-nuclear-reactors-151222131134565.html; Press. TV, Russian–Iranian business forum paves the way for investment cooperation, www.presstv.ir/Detail/2016/05/25/467265/Russia-Iran-business-forum; Russia interested in investment in Iran oil, gas projects, *Islamic Republic News Agency*, July 30, 2017, www8.irna.ir/en/News/82168725/; Sputnik, Moscow supports Tehran's willing to become SCO joint member, Russian President Vladimir Putin said in an interview, https://sputniknews.com/world/201608051043962974-russia-iran-sco-putin/; RNS, KAMAZ will supply to Iran in 2017 300 Trucks, December 14, 2016, https://rns.online/industry/KamaZ-postavit-v-Iran-v-2017-godu-300-mashinokomplektov-2016-12-14/; GAZ Group will supply 900 buses to Iran, www.rbc.ru/business/27/12/2016/586257949a794718ec812a66#xtor=AL-%5Binternal_traffic%5D--%5Brss.rbc.ru%5D-%5Btop_stories_brief_news%5D; Kommersant, VEB adjusted Iran's first deal [ВЭБ подогнал Ирану первую сделку], *Kommersant*, February 10, 2016, https://www.kommersant.ru/doc/2912481; TAAS, Iran's demand for Russian weaponry is estimated at $10 bln, *TASS Russian News Agency*, November 14, 2016, tass.com/defense/912141; Agreement for Purchasing of Sukhoi, *Fars News Agency*, www.farsnews.com/newstext.php?nn=13941120001820; TAAS, Iran plans negotiations with Russia on purchase of Sukhoi fighters, *TASS Russian News Agency*, August 22, 2016, tass.com/defense/895279; Iran, Russia Inching Closer to Su-30 Fighter Jet Deal, December 01, 2016, thediplomat.com/2016/12/iran-russia-inching-closer-to-su-30-fighter-jet-deal/; Telegraph, Roland Oliphant, Russia may sell Iran $10 billion worth of tanks and jets in new arms deal, *The Telegraph*, November 14, 2016, www.telegraph.co.uk/news/2016/11/14/russia-may-sell-iran-10-billion-worth-of-tanks-and-jets-in-new-a/.

[28] Forbes, When U.S. Sanctions Backfire: Russia And Iran Partner Up In $2.5B Deal, *Forbes*, August 1, 2017, https://www.forbes.com/sites/wadeshepard/2017/08/01/russia-and-iran-partner-up-as-the-us-turns-its-back/#403474165f63.

Much as the economy played a large part in Iran's pivot to the East, both countries have found a common interest in defending the Assad rule in Syria.

Laying the Foundations for a Russian–Iranian Condominium in Syria

Moscow's relations with the Assad dynasty preceded Vladimir Putin's plan to restore Russia's international prestige. The former Soviet Union had regarded Syria under Assad as its key asset in the Middle East, not least because its relations with Egypt had cooled in the 1980s. On October 8, 1980, the two countries signed the Treaty of Friendship and Cooperation which obligated Moscow to defend Syria in case of a third-party attack. The Treaty reaffirmed the Soviet lease of the naval facility in Tartus, the only Mediterranean port that serviced the Soviet Black Sea fleet. After the collapse of the Soviet Union relations unraveled because of Russia's political and economic difficulties, but President Putin forgave 80 percent of the huge Syrian debt, restoring the Soviet-era links.

Russia's alliance with Syria strengthened considerably after the Lebanese Prime Minister Rafik Hariri was assassinated in 2005. Blamed for the assassination, Syria, which faced increased hostility from the both the West and the Arab states, had turned to Russia for protection. Concurrently, Bashar al-Assad looked to Mahmoud Ahmadinejad, whose own anti-Americanism offered a good diplomatic match. The hardline leadership in Tehran viewed Russia's diplomatic overtures towards Syria very positively because Russia's close relationship with Syria complemented its anti-American policy. Iran's 2006 agreement with Assad on bolstering Syria's defensive capacity against the United States and Israel mirrored the expansion of Russian arms contracts to Syria.

As Chap. 6 details, the "Arab Spring" caught Assad and his Iranian backers by surprise. Like the Iranians, the Russian were unhappy with the regime's harsh reaction to the civilian protests. On August 1, 2011, the Russian Foreign Ministry issued a statement condemning the violence, including the alleged killing of eight policemen by Assad's government. The statement beseeched the pro-Assad forces as well as the violent protesters to "exercise maximum restraint."[29]

[29] MFARF, "Russian MFA Press and Information Department Comment on the Situation in Syria," *Ministry of Foreign Affairs of the Russian Federation*, August 1, 2011, http://www.mid.ru/en/maps/sy/-/asset_publisher/9fcjSOwMERcf/content/id/198682.

What mattered most to Ahmadinejad and the Islamic Revolutionary Guards Corps (IRGC) and Quds Force (QF), however, was the Russian diplomatic position in support of Syria. At the end of May 2011, the Russian Foreign Minister Sergei Lavrov rejected the United Nations involvement in Syria because "the situation doesn't present a threat to international peace and security." He added that Syria was a critical country in the Middle East and destabilizing it would have repercussions far beyond its borders. Lavrov also asserted to the world that Assad had made attempts at major reform. Starting in October 2011, Russia, a permanent member of the United Nations Security Council (UNSC) had vetoed the American-led motions to remove the Syrian president and impose sanctions on his regime. On November 1, Lavrov told a Russia–Gulf ministerial meeting that Russia would oppose the recent proposal for a no-fly zone in Syria as, in Russia's view, similar to the no-fly zone in Libya, it had been used to "support one side in a civil war."[30]

Since Russia played the diplomatic "bad cop," the IRGC and even the regular military felt free to threaten all those who considered replacing Assad. They stated that Tehran would not allow foreign powers to force Assad from power. Deputy Chief of Staff of the Iranian Armed Forces Brigadier General Massoud Jazayeri warned the West and the Arab countries to expect a "decisive hit" should they interfere. As he put it, "the enemy is not able to move in Syria, and the Syrian people and its friends will not allow this. No Syrian friend and no great resistance front have entered the arena yet, and if this happens, it will decisively hit the enemy, particularly the hateful Arabs.[31] Although Jazayeri did not mention Russia by name, it was quite clear that he counted Putin among Syria's good friends.

But the Russian diplomatic efforts were not enough to save the Assad government, which was losing ground to a coalition of rebel forces as documented in Chap. 6. To make matters worse for the Iranians, Moscow seemed to have turned its back on Assad. Many among the Russian ruling elite were privately unhappy with Assad's violent response to the Arab Spring demonstrations. The 2012 Houla massacre in which government forces killed civilians elicited particularly negative sentiments in Moscow.

[30] Henry Meyer et al., "Russia Warns U.S., NATO Against Military Aid to Syria Protests After Libya", *Bloomberg*, June 2, 2011; MEO, Russia will not allow Libya-style military intervention in Syria", *Middle East Online*, November 1, 2011, www.middle-east-online.com/english/?id=48833.

[31] FNA, Entering of Syria's Friends will hit Enemies Decisively, *Fars News Agency*, July 24, 2012, www.farsnews.com/newstext.php?nn=13910503000560.

On June 11, 2013, Putin acknowledged that Assad had not undertaken any "reforms" to remedy the situation that had led to the crisis. He told the Russian state media that: "Syria as a country was rife for change. And the government of Syria should have felt that in due time and should have undertaken some reform. Had they done that, what we're seeing in Syria today would have never happened."[32]

Putin's statement could, therefore, be taken as a nod to these critical factions within the Russian government. Most probably though, it was a signal to Saudi Arabia and the Sunni Arab countries in the Middle East. As noted in Chap. 7, Saudi Arabia took the lead in trying to depose the Assad regime and Moscow, anxious not to alienate such an important player, had to demonstrate that there were limits to Moscow's partnership with Damascus. It did not help Russia when, on August 21, Assad launched a chemical attack on civilians in Ghouta, a suburb of Damascus where more than a thousand people died, according to some estimates.

On its face, Putin's statement did little to harm the Russia–Iran relationship, but the hardliners were worried about a possible Russia–Saudi rapprochement. Things took a turn for the worse when Moscow canceled a diplomatic trip to Iran scheduled for August 2013. The IRGC leaders concluded that there was a commitment gap between Russia and Iran in Syria. Some commanders were reticent to partner with Russia because of skepticism about Moscow's long-term commitment to supporting the Assad cause. Observers noted that the Russians wanted to preserve a reformed Syrian state, rather than preserve Assad and his unpopular Alawite cohorts.[33]

Fortunately for the IRGC-QF, Putin's decision to annex the Crimea in 2014 changed the calculus in Moscow. After President Obama denounced Moscow and the West imposed sanctions on Russia, Putin sought ways to expand his Middle East presence and create trouble for the Americans to boot. Crimea (and thus Ukraine), the Straits (hence Turkey) and the eastern Mediterranean (Syria, Lebanon, Palestine) became, from the Russian

[32] Richard Spencer, Vladimir Putin admits Bashar al-Assad responsible for Syrian uprising". *The Daily Telegraph*, June 11, 2013, www.telegraph.co.uk/news/worldnews/middleeast/syria/10114247/Vladimir-Putin-admits-Bashar-al-Assad-responsible-for-Syrian-uprising.html.

[33] Dmitri Trenin, Russia and Iran: Historic Mistrust and Contemporary Partnership, August 18, 2016, carnegie.ru/2016/08/18/russia-and-iran-historic-mistrust-and-contemporary-partnership-pub-64365.

perspective, one long front.[34] Extending military assistance to Assad who was losing ground to the U.S. and Saudi supported rebels was an optimal way to achieve this goal. As will be elaborated in Chap. 6, the QF chief Qassem Suleimani traveled to Moscow several times to meet with Putin and his military commanders to discuss ways to help Assad. An article in the pro-Hezbollah newspaper *al-Akhbar* on September 22, 2014, revealed that Russia had entered an agreement with Iran, Syria, and Iraq, and Hezbollah. The paper noted that the so-called 4+1 alliance "includes administrative mechanisms for cooperation on [the issues of] politics and intelligence and [for] military [cooperation] on the battlefield in several parts of the Middle East, primarily in Syria and Iraq." The paper also claimed that an advanced party of Russian military arrived in Syria to assess the battlefield needs. Farsi media indicated that Russia agreed to furnish the Syrian army with "special" weapons.[35]

During the Kremlin meeting, Suleimani suggested Russia adheres to his original "binary option" plan. Discussed in greater detail in Chap. 6, the scheme, first proposed by Suleimani to Assad in 2011, entailed creating a radical jihadist group. The QF chief, drawing from his vast experience in Iraq, believed that the presence of an extremist group would make Assad look like a lesser evil in the eyes of his people and the West. He was partially vindicated when the disparate Syrian jihadist joined al-Qaeda in Iraq to form the notorious ISIS.

The Russian–Iranian Strategy of Preserving the "Binary Option:" Fighting the Anti-Assad Insurgents

When the Russians deployed to Syria in September 2015, they gave the weakened Syrian Army, the Hezbollah, and the IRGC-QF a timely relief. The Russians arrived with an impressive array of air and land power. The air force committed some of its most lethal weapons, including two Mi-28s,

[34] Audrey L. Altstadt "Putin's Middle East Triangle," Kenan Cable No. 23, *Wilson Center*, June 19, 2017, https://www.wilsoncenter.org/publication/kennan-cable-no23-putins-middle-east-triangle.

[35] Pro-Hezbollah daily says party in Syria pact with Russia, *NOW*, September 22, 2015, https://now.mmedia.me/lb/en/NewsReports/565949-pro-hezbollah-daily-says-party-in-syria-pact-with-russia; YJC, Suleimani-Putin Meeting in Moscow: Russia's Key to Attack Syria, *Yong JournalistsClub*, www.yjc.ir/fa/news/5350142/ دیدار-سه-ساعته-پوتین-و-سردار-سلیمانی-در-مسکو-کلید-حمله-روسیه-به-سوریه; Shoayb Bahman, Suleimani-Puting Agreement (Tehran, Andisheh Sazan-e Noor) 2016.

two-seat anti-tank attack helicopters, capable of operating in all weather conditions, day or night. Moscow also sent 12 × Su-24, supersonic, all-weather attack aircraft/interdictors; 12 × Su-25; 12 × Mi24, a large helicopter gunship; 4 × Su-30, all-weather multirole fighter. Air-to-air and air-to-surface deep interdiction missions, 8 × SU-34, twin-engine, twin-seat strike fighters designed to destroy targets at sea, air and on land, including small and mobile ones. 14 × Tu-22 M, supersonic, long-range strategic and maritime strike bombers, 6 × TU-95, large, four-engine strategic bomber and missile carriers, 5 × Tu-160, supersonic heavy strategic bombers, 4 × Su-35 twin-engine, highly maneuverable fighter jets were also included. One × Il-20 and one × Tu-214 reconnaissance aircraft, 4 × Mi-8, and 2 × Ka-52 'Alligator' attack helicopters completed the arsenal.[36]

A warship armed with cruise missiles has joined the Russian battle group off the coast of Syria as part of Moscow's response to U.S. cruise missile strikes on the Syrian airbase of Shayrat. The Russian aircraft carrier Kuznetsov—which carried more than 50 aircraft—escorted by seven other Russian ships have also deployed in Syria to form Russia's largest naval deployment since the end of Cold War. The other vessels accompanying the carrier included the nuclear-powered battle cruiser, *Peter the Great*. The deployment also included two large anti-submarine warships as well as four support vessels. According to a Norwegian Lockheed P-3 Orion reconnaissance plane, MiG-29 Fulcrum jets and combat helicopters were visible on the deck of the carrier.[37]

Ground forces completed Moscow's deployment, including several artillery units and 7 × T-90s tanks, Russia's most advanced tanks, which were stationed near the Syrian airbase near Latakia. Moscow has also fielded a land-based cruise missile, Howitzer Batteries code name "Gabatr 120th ABR 2A65 Msta B". Paratrooper and special forces comprised Russian ground forces. Military experts speculated that this rather excessive show of force went beyond helping Assad. Syria became a proving ground for a new generation of weapons that Russia developed to update its antiquated arsenal and a laboratory for new military strategies in which special operation forces played a large part. They also argued that the

[36]The Lethal Russian Air Power Deployed in Syria, *Forces Network News*, September 23, 2016, www.forces.net/news/tri-service/lethal-russian-air-power-deployed-syria.

[37]Alec Luhan, Russia sends warship to battlegroup off Syrian coast, April 8, 2017, https://www.theguardian.com/world/2017/apr/08/russia-sends-warship-syrian-coast.

Russians planned to use some of the hardware to build a permanent defense position in areas that would be later designated as de-escalation zones.[38]

While Moscow created a narrative of fighting ISIS, the Russian strategy, guided by the QF, targeted mostly non-ISIS groups. A report by the Atlantic Council which used digital forensics and social media intelligence (OSSMINT) demonstrated that the vast majority of Russian aerial and ground attacks in Syria targeted anti-Assad forces, such as the US-backed Free Syria Army (FSA) and the Kurdish fighters. To help the Assad regime, the Russians bombed civilian facilities, namely, mosques, hospitals, and schools, in areas occupied by the Assad opponents.[39]

In almost all cases, the strikes followed the same pattern. The Syrian, Iranian and Russian state media reported initial strikes against what was described as "ISIS dens" such as al-Rastan, Al-Latamneh, Hama, Talbeisa, al-Zaafran, al-Tolol al-Humr, Aydon, Deir Fol, and the area surrounding Salmia in the central region in Syria. However, there was no known ISIS presence in these locations, which served anti-Assad opposition groups. According to the Atlantic Council report, out of the 30 Russian claims of ISIS hits, only one was confirmed as such.[40]

Worse, the Russian Air Force appears to have used banned RBK-500 cluster munitions during its Syrian campaign. In December 2015, Human Rights Watch and Amnesty International published reports detailing the use of cluster munitions by Russian forces in Syria. These and similar analyses left little doubt that Assad was the main beneficiary of the Russian air strikes, and the Western supported fighters and civilians were the main victims.[41]

Using the Russian bombing as a shield, Assad's forces were gradually able to advance, reversing their earlier territorial losses. In the Aleppo

[38] Alec Luhn, Russia sends artillery and tanks to Syria as part of continued military buildup, *The Guardian*, September 14, 2015, https://www.theguardian.com/world/2015/sep/14/russia-sends-artillery-and-tanks-to-syria-as-part-of-continued-military-buildup; Reuters, Army map suggests presence of Russian artillery unit in central Syria, November 18, 2015, www.reuters.com/article/us-mideast-crisis-syria-russia-military-idUSKCN0T72VW20151118; Paul Iddon, For Russia, Syria Is a Proving Ground for High-Tech Weapons, *War is Boring*, January 4, 2017, https://warisboring.com/for-russia-syria-is-a-proving-ground-for-high-tech-weapons/.

[39] The Guardian, More than 90%' of Russian airstrikes in Syria have not targeted Isis, U.S. says, *The Guardian*, October 7, 2015, https://www.theguardian.com/world/2015/oct/07/russia-airstrikes-syria-not-targetting-isis.

[40] Maksymilian Czuperski et al., Distract, Deceive, Destroy: Putin At War in Syria, *Atlantic Council of the United States*, April 2016.

[41] Ibid.

region, pro-government forces expanded control southwest of Aleppo city, and cutting off the non-ISIS rebel supply route to the north with Turkey. Strikingly, these advances appear to have been backed by Russian regular army units firing 152 mm-caliber Msta-B howitzers—despite Russian claims that it had no troops fighting on the ground.[42]

Apparently, the Atlantic Council findings did not come as a surprise to the Obama administration. As early as October 2015, a State Department spokesperson stated: "over 90 percent of the strikes had targeted neither ISIS nor al-Qaeda affiliates." Shortly after, President Obama noted that "Russia's strikes against the moderate opposition only bolster the Assad regime." But the administration, fresh from its successful nuclear deal with Iran, was reluctant to publicly acknowledge that Russia was working in tandem with the Revolutionary Guards to promote the "binary option." The White House was also indebted to Moscow, which was helping Obama to solve his Ghouta dilemma. Unwilling to retaliate even after Syria crossed the "red line," the president's depiction of a chemical attack on civilians, Russia's offer to supervise the removal, and the destruction of all Syrian chemical weapons stockpile.[43]

By any measure, the massive Russian intervention was critical in reversing the gains made by the non-jihadist anti-Assad opposition. By selective targeting of groups, the Russians were essentially carving a Russian–Iranian condominium. After years under rebel control, Aleppo, the country's industrial and economic capital, had fallen to pro-government troops in December 2016. The loss of Aleppo left Syria's opposition at its weakest points since the beginning of the uprising in 2011. As the war progressed, the Russian helped the Iranians carve new strategically important territory, a topic that is discussed more fully in Chap. 6.[44]

After cashing on the "binary option" victory, Russia worked hard to reward Iran in the diplomatic arena. The official United Nation's track offered an initial opening. In January 2014, the Geneva Conference on Syria, organized by Lakhdar Brahimi, the UN envoy to Syria, commenced a collaboration with the United States and Russia. In addition to representatives from the Syrian government and opposition, delegates from

[42] Ibid.

[43] Ibid., p. 12.

[44] Nick Robins-Early, What The Fall Of Aleppo Means For Syria's Civil War, *Huffington Post*, December 15, 2016, www.huffingtonpost.com/entry/syria-aleppo-war_us_58505b11e4b092f08685dfeb; Aron Lund, The fall of eastern Aleppo, December 13, 2016, https://www.irinnews.org/analysis/2016/12/13/fall-eastern-aleppo.

Canada, China, France, Germany, Russia, Saudi Arabia, Turkey, United Kingdom, United States, and Iran joined in to discuss a transitional government. Washington demanded that Assad step down before the formation of a transition government, an anathema to both Russia and Iran. The talks yielded no tangible results and Brahimi resigned.[45]

But Iran gained another diplomatic platform when, on May 22, 2015, the president Nursultan Nazarbayev convened a conference in Astana, the capital of Kazakhstan. Several Syrian opposition groups were in attendance. Unlike in Geneva, Russia and Iran played a major part in the meeting. Following intense and complicated negotiations, the participants issued a joint statement on May 28 titled "Astana Declaration for a Political Solution in Syria." The document stressed the need for "a gradual political transition of power through a provisional government of national accord."[46]

During the second Astana meeting on October 2–4, 2015, the opposition groups adopted a declaration calling for the parliamentary election to be held under the supervision of the international community in 2016. But the Russian airstrikes caused heated discussions and the parties failed to reach a consensus. Eventually, the text of the declaration did not include an evaluation of the Russian Air Force military operation against ISIS launched in Syria at the request by President Bashar al-Assad.[47]

[45] UNP, Final Communiqué of the Action Group for Syria (Geneva Communiqué), June 30, 2012, peacemaker.un.org/node/1581; PS, "Syria, a civil, sectarian and proxy war" Interview with Lakhdar Brahimi, *Politically Speaking*, July 2013.

The other participants were Algeria, Brazil, Denmark, Egypt, India, Indonesia, Iraq, Italy, Japan, Jordan, Kuwait, Lebanon, Morocco, Norway, Oman, Qatar, South Africa, Spain, Sweden, Switzerland, United Arab Emirates; Aljazeera, Syria mediator Brahimi announces resignation, *Aljazeera*, May 14, 2014, www.aljazeera.com/news/middleeast/2014/05/syria-peace-envoy-brahimi-201451315191 8573244.html.

[46] Malika Orazgaliyeva, Syrian Opposition Meeting in Astana Makes Tentative Headway in Finding Way Forward for Peace Process, *The Astana Times*, May 29, 2015, astanatimes.com/2015/05/syrian-opposition-meeting-in-astana-makes-tentative-headway-in-finding-way-forward-for-peace-process/; Vitaly Naumkin, Russia mulls strategy to solve Syrian conflict, June 3, 2015, www.al-monitor.com/pulse/originals/2015/06/russia-syria-conflict-strategy-moscow-meetings.html.

[47] Regnum, Syrian opposition held a conference in Astana (Сирийская оппозиция провела конференцию в Астане), *Regnum.Ru*, October 4, 2015, https://regnum.ru/news/polit/1984197.html; Vestnik, Second round of Syrian opposition talks begins in Astana, October 2, 2015, vestnikkavkaza.net/news/Second-round-of-Syrian-opposition-talks-begins-in-Astana.html; The Syrian opposition signed the final declaration of the conference in Astana, *RIA Novosti*, October 4, 2015, https://ria.ru/world/20151004/1296432701.html; Sputnik, Syrian

Russia displayed its diplomatic prowess again when 17 nations met in Vienna on November 12, 2015, to adopt a timeline for a transition plan in Syria. While Russia and Iran had sided with Assad, the U.S. and its regional allies had insisted upon his removal. The resolution called for "credible, inclusive and nonsectarian governance" within 6 months and "free and fair elections, pursuant to the new constitution" within 18 months, and a 2-year timeline to create a unity government and hold elections. Although Russia voted for the resolution, Iran and Assad convinced Moscow not to help with its implementation.[48]

With the Obama administration reluctant to engage in the Syrian civil war and the European Union reduced to periodical but rather meaningless declarations in favor of settlement, Iran could count on Russian diplomatic opposition to Assad's removal. But as the preceding chapter indicated, both the normalizers and the hardliners in Tehran were worried about President Donald J. Trump.

The Russian–Iranian Relations in the Trump Era: Future Prospect for the Condominium

As a candidate, President Trump chastised his predecessor for abandoning the Syrian people to the "tyrant" Bashar al-Assad. Trump mocked President Obama for retreating on his "red line" promise to retaliate for the regime's use of chemical weapons. In fact, when the Syrian Air Force dropped sarin gas on the residence of Khan Shykhun in the rebel-held province of Idlib on April 4, 2017, the Americans struck the Shayrat airbase implicated in the chemical attack with 55 Tomahawk cruise missiles.

Opposition Signs Final Declaration After Marathon Talks in Astana, October 4, 2015, https://sputniknews.com/politics/201510041027986345-syrian-opposition-final-declaration/.

[48] Jonathan Tirone, Syrian Transition Plan Reached by U.S., Russia in Vienna, *Bloomberg*, November 14, 2015, https://www.bloomberg.com/amp/news/articles/2015-11-14/syrian-transition-plan-achieved-by-u-s-allies-kerry-says; Ceasefire and transition plan for Syria under discussion at Vienna talks, www.dw.com/en/ceasefire-and-transition-plan-for-syria-under-discussion-at-vienna-talks/a-18817325; Security Council Unanimously Adopts Resolution 2254 (2015), Endorsing Road Map for Peace Process in Syria, Setting Timetable for Talks, www.un.org/press/en/2015/sc12171.doc.htm; Security Council Unanimously Adopts Resolution 2254 (2015), Endorsing Road Map for Peace Process in Syria, Setting Timetable for Talks, SC/12171, Security Council 7588TH MEETING (PM), 18 December 2015; Alexey Firsov, A shot down plane, a "drunk" boy, old Tuleyev., [Сбитый самолет, «пьяный» мальчик, старожил Тулеев.], pltf.ru/2017/06/23/povestka-nedeli-23-june/.

On February 28, 2017, after denouncing Trump's move, the Russians, along with China, vetoed a UNSC resolution authorizing sanctions against the Syrian government for its use of chemical weapons. In explaining the decision, Vladimir Safronkov, the Russian ambassador to the United Nations, suggested that the evidence was uncorroborated and came from "suspicious eyewitness accounts … armed opponents, sympathetic [nongovernmental organizations], media and also the so-called Friends of Syria."[49]

In retrospect, Russia's robust defense of Assad, which was echoed in Tehran, indicated Moscow's ambitious effort to end the civil war in Syria on terms that would preserve both the Assad regime and the condominium project. On May 4, 2017, Iran, Russia along with Turkey, signed an agreement in Astana, to create the so-called de-escalation or "safe zones." The deal covered four of the Syria's most congested areas: the Idlib province, mainly controlled by jihadist and other rebel groups; Rastan and Talbiseh enclaves in the Homs province; Eastern Ghouta, north of Damascus; and southern Syria along the Jordanian and the Israeli border, including the Deraa and Quneitra provinces.[50]

Originally suggested by Qassem Suleimani, the de-escalation of zones where more than 2.5 million people lived was a smart strategy to attain a number of goals. It minimized the bloodshed and the concomitant attention of the United Nations. It legitimized the role of Iran as a major player in Syria and virtually assured the territorial gains of Russia around Tarsus and Latakia on the one hand, and the Iranian land bridge, discussed in more detail in Chap. 6. Equally important, the Astana agreement presented the United States with a fait accompli that left little room for a robust military response.[51]

[49] Russia, China veto at U.N. on Syria chemical weapons is 'outrageous,' U.S. says, *The Washington Post*, February 28, 2017, https://www.washingtonpost.com/world/national-security/russia-china-veto-at-un-on-syria-chemical-weapons-is-outrageous-us-says/2017/02/28/c69adcf4-fdeb-11e6-99b4-9e613afeb09f_story.html?utm_term=.661523f59a8e.

[50] PressTV, The Deal to Establish Safe Zones in Syria is Reached, *PressTV*, May 6, 2017, presstv.ir/DetailFr/2017/05/06/520730/Syria-agreement-deescalation-zones-Russia; Josie Ensor, Turkey, Russia and Iran sign 'safe zones' deal for Syria, *The Telegraph*, May 4, 2017, www.telegraph.co.uk/news/2017/05/04/turkey-russia-iran-sign-safe-zones-deal-syria/.

[51] PressTV, The Deal Establish Safe Zones in Syria is Reached, *PressTV*, May 6, 2017, presstv.ir/DetailFr/2017/05/06/520730/Syria-agreement-deescalation-zones-Russia; Josie Ensor, Turkey, Russia and Iran sign 'safe zones' deal for Syria, *The Telegraph*, May 4, 2017, www.telegraph.co.uk/news/2017/05/04/turkey-russia-iran-sign-safe-zones-deal-syria/.

An analysis of American military actions in the wake of the de-escalation deal indicated a rather circumspect attitude in defending the pro-Western forces from the Russian–Iranian encroachments. The administration tried to disrupt the continuity of the land bridge by seizing several strategic positions from ISIS, including al-Tanf where an American garrison trained anti-Assad fighters. The Americans declared the area around al-Tanf a de-escalation zone; that is, a 55-kilometer no-go zone. When, on June 6, Suleimani ordered 60 fighters armed with tanks and anti-aircraft weapons to enter the territory, the American bombed the convoy after it failed to heed a warning to leave the area. Two days later, an American jet shot down an IRGC Shahed-129 drone that crossed into the exclusion zone.[52]

On June 18, 2017, the U.S. military shot down a Syrian plane near a town outside Raqqa controlled by the Syrian Democratic Forces (SDF), an American ally. According to a Pentagon spokesman, the Syrian jet had dropped bombs near an American coalition force engaged in anti-ISIS operations. Joseph Dunford, chairman of the Joint Chiefs of Staff, justified the action, stating that the Americans needed to protect their personnel. Previously he explained that it "was a force protection strike. Our commanders on the ground felt like they were threatened at that point. And their rules of engagement allow them to do that."[53]

But on other occasions, the administration showed little resolve in challenging the Russians, and by extension, their Iranian partner: among them being the failure to prevent the IRGC-trained and Russian-supported Syrian 5 Corps from crossing into the east bank of the Euphrates river. Moscow ignored Washington's warning that such a move would put the Iranian and Syrian units forces in conflict with the American-backed Syrian Democratic Forces (SDF) and furnished pontoon bridges for the crossing. Conservative critics of the Pentagon warned that if the U.S. wants to support "local actors such as the SDF…to contain Iran and its

[52] Paul Mccleary, U.S. Bombs Iranian Militia in Syria, As Fight For Raqqa Begins, *Foreign Policy*, June 6, 2017, webcache.googleusercontent.com/search?q=cache:fKUE86qc6JMJ:foreignpolicy.com/2017/06/06/u-s-bombs-iranian-militia-in-syria-as-fight-for-raqqa-begins/+&cd=1&hl=en&ct=clnk&gl=tr.

[53] USDOD, Department of Defense Press Briefing by Secretary Mattis, General Dunford and Special Envoy McGurk on the Campaign to Defeat ISIS in the Pentagon Press Briefing Room, https://www.defense.gov/News/Transcripts/Transcript-View/Article/1188225/department-of-defense-press-briefing-by-secretary-mattis-general-dunford-and-sp/; Nabih Bulos and W.J. Hennigan, U.S. warplane shoots down Syrian jet near Raqqah, *Los Angeles Times*, June 18, 2017, www.latimes.com/world/middleeast/la-fg-syria-us-jet-20170619-story.html.

allies, then Washington needs to recognize that Tehran and Assad are directly challenging this goal with the help of Russian airpower."[54]

A more assertive American posture could probably help to limit the condominium project, but the pivot to the East is firmly entrenched on the Iranian side. How much Russia would invest in the alliance is not entirely clear, however. Some of Moscow's interests, such as maintaining good relations with the Sunni players—Saudi Arabia, the Gulf Cooperation Council, and Egypt. As Fyodor Lukyanov, the editor-in-chief of *Russia in Global Affairs* and chairman of the Presidium of the Council on Foreign and Defense Policy, put it, the "deliberate support of the Shiite" against the Sunnis would be "suicidal."[55] Moscow, which has good relations with Israel, is hardly supportive of Iran's anti-Israel crusade. On a tactical level, Russia would need to curb possible clashes between the IRGC-backed militias and Hezbollah on the one side and the Israeli forces on the other, a concern discussed in Chap. 9.

Conclusion

Although the regime was suspicious of communism, it avoided antagonizing the former Soviet Union, a posture that paid off handsomely. In addition to aiding Iran with its nascent nuclear program, Russia acted as a sanction-busting "black knight," that mitigated some of the effects of the draconian sanctions imposed on Iran to force the termination of its nuclear program. President Mahmoud Ahmadinejad tried to utilize these ties to execute the so-called pivot to the East, a policy which sought to relocate Iran's economic ties to the Shanghai Economic Cooperation (SEC), a political-economic structure akin the Western-oriented institutions like the World Bank and IMF, among others.

Both the moderates and the hardliners have welcomed Putin's strong defense of the JCPOA in the face of American pressures. The Islamic Republic also shares with Moscow two meta-goals of weakening the EU and undermining NATO.

Yet the normalizers and the Principalists differ about the degree to which Iran should pivot toward Russia. Rouhani and his followers feel that there should be a balance between the Western and Eastern orientation,

[54] Andrew J. Tabler, Russia Crosses the Euphanes. Implications. *The Washington Institute*, September 18, 2017.

[55] Paul J. Saunders, "Rethinking Russia's Mideast Ambitions," *Al Monitor*, July 3, 2017.

even as America becomes more isolationist. The Principalists are in favor of a sharp tilt toward Moscow, not least because of their strong military and nuclear cooperation. With both countries mobilized to help Bashar al-Assad retain power, the war in Syria brought Iran and Russia closer. The close collaboration between the Guards and Russia in Syria had further bolstered the position of the hardliners.

References

A Lot of Work to Do': What's on the Agenda of the Putin–Rouhani Meeting? https://sputniknews.com/world/201703271052004318-russia-iran-putin-rouhani-visit/.

Agreement for Purchasing of Sukhoi, *Fars News Agency*, www.farsnews.com/newstext.php?nn=13941120001820.

Alec Luhan, Russia sends warship to battlegroup off Syrian coast, April 8, 2017, https://www.theguardian.com/world/2017/apr/08/russia-sends-warship-syrian-coast.

Alec Luhn, Russia sends artillery and tanks to Syria as part of continued military buildup, *The Guardian*, September 14, 2015, https://www.theguardian.com/world/2015/sep/14/russia-sends-artillery-and-tanks-to-syria-as-part-of-continued-military-buildup.

Alexey Firsov, A shot down plane, a "drunk" boy, old Tuleyev. [Сбитый самолет, «пьяный» мальчик, старожил Тулеев.], pltf.ru/2017/06/23/povestka-nedeli-23-june/.

Ali Akbar Velayati, Eastern Partnership: The status of the East and West in the foreign policy of the Islamic Republic of Iran Hamshahri Diplomatic, Vol. 1, No 3.

Ali Jalali, "The Strategic Partnership of Russia and Iran," *Parameters*, Winter 2001–2002.

Alireza Sheikhattar, Accompanying Diplomacy and Economics: The Place of Economic Diplomacy in Regional Cooperation of the Islamic Republic of Iran, *Hamshahri Diplomatic*, Vol. 1, No 3. 2006.

Aljazeera, Russia to start building Iran nuclear reactors, www.aljazeera.com/news/2015/12/russia-start-building-iran-nuclear-reactors-151222131134565.html.

Aljazeera, Syria mediator Brahimi announces resignation, *Aljazeera*, May 14, 2014, www.aljazeera.com/news/middleeast/2014/05/syria-peace-envoy-brahimi-201451315191857324 4.html.

Andrew J. Tabler, Russia Crosses the Euphanes. Implications, *The Washington Institute*, September 18, 2017.

Andrew Tortman, Vladimir Putin signs historic $20bn oil deal with Iran to bypass Western sanctions, *The Telegragh*, August 6, 2014, www.telegraph.co.uk/finance/newsbysector/energy/oilandgas/11014604/Vladimir-Putin-signs-historic-20bn-oil-deal-with-Iran-to-bypass-Western-sanctions.html.

Ariel Cohen, Countering Russian-Iranian Military Cooperation, *The Heritage Foundation*, April 5, 2001, www.heritage.org/europe/report/countering-russian-iranian-military-cooperation.

Aron Lund, The fall of eastern Aleppo, December 13, 2016, https://www.irinnews.org/analysis/2016/12/13/fall-eastern-aleppo.

Arthur Blinov, Iran in the crossroads of geopolitics (Иран в перекрестье геополитики), www.ng.ru/courier/2006-01-16/14_iran.html?id_user=Y.

Audrey L. Altstadt "Putin's Middle East Triangle," Kenan Cable No. 23, *Wilson Center*, June 19, 2017, https://www.wilsoncenter.org/publication/kennan-cable-no23-putins-middle-east-triangle.

Audrey L. Altstadt, Kennan Cable No.23: Putin's Middle East Triangle, *Wilson Center*, June 19, 2017, https://www.wilsoncenter.org/publication/kennan-cable-no23-putins-middle-east-triangle.

Cameron R. Hume, The United Nations, Iran, and Iraq: How Peacemaking Changed (Indiana University Press, 1994).

Ceasefire and transition plan for Syria under discussion at Vienna talks, www.dw.com/en/ceasefire-and-transition-plan-for-syria-under-discussion-at-vienna-talks/a-18817325.

Debalina Ghoshal, Russia-Iran: Why is the S-300 Sale Significant?, *Institute of Peace and Conflict Studies*, June 19, 2015, www.ipcs.org/article/iran/russia-iran-why-is-the-s-300-sale-significant-4891.html.

Diesel-electric submarines of Project 877 "Paltus" (Дизель-электрические подводные лодки проекта 877 «Палтус»), *Boehhoe*, February 5, 2013, https://topwar.ru/23840-dizel-elektricheskie-podvodnye-lodki-proekta-877-paltus.html.

Dmitri Trenin, Russia and Iran: Historic Mistrust and Contemporary Partnership, August 18, 2016, carnegie.ru/2016/08/18/russia-and-iran-historic-mistrust-and-contemporary-partnership-pub-64365.

Domestic Economy, Tehran, Moscow to Sign Investment Guarantee Deal, November 16, 2015, https://financialtribune.com/articles/domestic-economy/30319/tehran-moscow-to-sign-investment-guarantee-deal.

Donyaye Eghtesad, The First Step for Creation of Gas OPEC, *Donyaye Eghtesad*, February 3, 2007, donya-e-eqtesad.com/news/665687.

Farhad Rezaei, Iran's Nuclear Program: A Study in Proliferation and Rollback (New York: Palgrave Macmillan, 2016).

FNA, Entering of Syria's Friends will hit Enemies Decisively, *Fars News Agency*, July 24, 2012, www.farsnews.com/newstext.php?nn=13910503000560.

FNA, If we look at the West, we need to take a begging bowl, *Fars News Agency*, www.farsnews.com/13960207000849.

FNA, Vladimir Putin Soon to Sign Historic $20bln Oil Deal with Iran to Bypass Western Sanctions, *Fars News Agency*, August 6, 2014, en.farsnews.com/newstext.aspx?nn=13930515000328.

Forbes, When U.S. Sanctions Backfire: Russia and Iran Partner Up In $2.5B Deal, *Forbes*, August 1, 2017, https://www.forbes.com/sites/wadeshepard/2017/08/01/russia-and-iran-partner-up-as-the-us-turns-its-back/#403474165f63.

GAZ Group will supply 900 buses to Iran, www.rbc.ru/business/27/12/2016/586257949a794718ec812a66#xtor=AL-%5Binternal_traffic%5D--%5Brss.rbc.ru%5D-%5Btop_stories_brief_news%5D.

Gevorg Mirzayan, We have agreements for 40 billion, [У нас соглашения на 40 миллиардов], *Expert Online*, expert.ru/2016/07/17/u-nas-soglasheniya-na-40-milliardov/.

Henry Meyer et al. "Russia Warns U.S., NATO Against Military Aid to Syria Protests After Libya", *Bloomberg*, June 2, 2011.

Hibbs, "German officials say governments may not let Bushehr be finished," *Nucleonics Week*, Vol. 32, No. 6 (1991).

Iran Newspaper, Strategy of "Looking to the East" in Rouhani Government, www.iran-newspaper.com/newspaper/BlockPrint/157208.

Iran Repairs the Five of the Former Iraqi MIG-29 (ИРАН ОТРЕМОНТИРОВАЛ ПЯТЬ БЫВШИХ ИРАКСКИХ МИГ-29), April 16, 2012, www.militaryparitet.com/perevodnie/data/ic_perevodnie/2741/%20https://topwar.ru/23840-dizel-elektricheskie-podvodnye-lodki-proekta-877-paltus.html.

Iran will give Russia priority in any industry it wants to invest in, April 21, 2016, https://in.rbth.com/news/2016/04/21/iran-will-give-russia-priority-in-any-industry-it-wants-to-invest-in_586933.

Iran, Russia Inching Closer to Su-30 Fighter Jet Deal December 01, 2016, thediplomat.com/2016/12/iran-russia-inching-closer-to-su-30-fighter-jet-deal/.

Iran's demand for Russian weaponry is estimated at $10 bln, *TASS Russian News Agency*, November 14, 2016, tass.com/defense/912141.

Iranian T.V Channel-2, "Iranian President Mahmoud Ahmadinejad: So What If Two U.N. Resolutions Were Issued against Us? America Cannot Cause Iran Real Damage," qouted by Middle East Media Research Institute MEMRI, January 23, 2007, http://www.memritv.org/clip_transcript/en/1369.htm.

Jonathan Tirone, Syrian Transition Plan Reached by U.S., Russia in Vienna, *Bloomberg*, November 14, 2015, https://www.bloomberg.com/amp/news/articles/2015-11-14/syrian-transition-plan-achieved-by-u-s-allies-kerry-says.

Josie Ensor, Turkey, Russia and Iran sign 'safe zones' deal for Syria, *The Telegraph*, May4,2017,www.telegraph.co.uk/news/2017/05/04/turkey-russia-iran-sign-safe-zones-deal-syria/.

Khamenei.ir, the 20-Year Perspective Document for Iran, farsi.khamenei.ir/special?id=4133.

Kommersant, Russia is localized in Iran, [Россия локализуется в Иране], *Kommersant*, https://www.kommersant.ru/doc/3170040.

Kommersant, VEB adjusted Iran's first deal [ВЭБ подогнал Ирану первую сделку], *Kommersant*, February 10, 2016, https://www.kommersant.ru/doc/2912481.

MacAskill, & Tait, "Europe proposes limited sanctions to halt Tehran's nuclear ambitions," *The Guardian*, April 11, 2006, http://www.theguardian.com/world/2006/apr/11/nuclear.eu.

Maksymilian Czuperski et al., Distract, Deceive, Destroy: Putin At War in Syria, *Atlantic Council of the United States*, April 2016.

Malika Orazgaliyeva, Syrian Opposition Meeting in Astana Makes Tentative Headway in Finding Way Forward for Peace Process, *The Astana Times*, May 29, 2015, astanatimes.com/2015/05/syrian-opposition-meeting-in-astana-makes-tentative-headway-in-finding-way-forward-for-peace-process/.

Mark N. Katz, Russian–Iranian Relations in the Ahmadinejad Era, *Middle East Journal*, Vol. 62, No. 2, Spring 2008.

Martin Sicker, The Bear and the Lion: Soviet Imperialism and Iran (New York: Praeger Publishers, 1988).

MEO, Russia will not allow Libya-style military intervention in Syria. *Middle East Online*, November 1, 2011, www.middle-east-online.com/english/?id=48833.

MFARF, "Russian MFA Press and Information Department Comment on the Situation in Syria," *Ministry of Foreign Affairs of the Russian Federation*, August 1, 2011, http://www.mid.ru/en/maps/sy/-/asset_publisher/9fcjSOwMERcf/content/id/198682.

Michael Eisenstadt, Russian Arms and Technology Transfers to Iran: Policy Challenges for the United States, *Arms Control Today* March 2001, https://www.armscontrol.org/print/807.

Michael Parks, "Iranian Gets Warm Soviet Welcome: Rafsanjani-Gorbachev Talks Point to Improved Relations," *Los Angeles Times*, June 21, 1981.

Mottaki, M, "Iran's Foreign Policy under President Ahmadinejad's," *Discourse An Iranian Quarterly* (Winter 2009).

Nabih Bulos and W.J. Hennigan, U.S. warplane shoots down Syrian jet near Raqqah, *Los Angeles Times*, June 18, 2017, www.latimes.com/world/middleeast/la-fg-syria-us-jet-20170619-story.html.

Nick Robins-Early, What The Fall Of Aleppo Means For Syria's Civil War, *Huffington Post*, December 15, 2016, www.huffingtonpost.com/entry/syria-aleppo-war_us_58505b11e4b092f08685dfeb.

Nikita Smagin, How Russia managed to double its exports to Iran in 2016, *Russia Beyond*, February 17, 2017, https://www.rbth.com/business/2017/02/17/russia-exports-iran-704108.

NIOC, OPEC Gas, from an understanding to an official organization, *National Iranian Oil Company*, www.nioc.ir/Portal/Home/ShowPage.aspx?Object=NEWS&ID=9d4809fc-88bd-45ea-96eb-74ef0c2e0ae0&LayoutID=d2b4e819-225e-417a-9564-dd14973277fc&CategoryID=5a83de0d-4c67-4b8a-a794-08e669faf33d.

Paul Iddon, For Russia, Syria Is a Proving Ground for High-Tech Weapons, *War is Boring*, January 4, 2017, https://warisboring.com/for-russia-syria-is-a-proving-ground-for-high-tech-weapons/.

Paul J. Saunders, "Rethinking Russia's Mideast Ambitions," *Al Monitor*, July 3, 2017.

Paul Mccleary, U.S. Bombs Iranian Militia in Syria, As Fight For Raqqa Begins, *Foreign Policy*, June 6, 2017, foreignpolicy.com/2017/06/06/u-s-bombs-iranian-militia-in-syria-as-fight-for-raqqa-begins/+&cd=1&hl=en&ct=clnk&gl=tr.

PressTV, Russian–Iranian business forum paves the way for investment cooperation, www.presstv.ir/Detail/2016/05/25/467265/Russia-Iran-business-forum.

PressTV, Iran president to pay visit to Russia for key talks, *PressTV*, www.presstv.ir/Detail/2017/03/24/515463/Iran-Russia-Hassan-Rouhani-Vladimir-Putin-Parviz-Esmaeili-Moscow.

PressTV, The Deal Establish Safe Zones in Syria is Reached, *PressTV*, May 6, 2017, presstv.ir/DetailFr/2017/05/06/520730/Syria-agreement-deescalation-zones-Russia.

Pro-Hezbollah daily says party in Syria pact with Russia, *NOW*, September 22, 2015, https://now.mmedia.me/lb/en/NewsReports/565949-pro-hezbollah-daily-says-party-in-syria-pact-with-russia.

PS, "Syria, a civil, sectarian and proxy war" Interview with Lakhdar Brahimi, *Politically Speaking*, July 2013.

Regnum, Syrian opposition held a conference in Astana (Сирийская оппозиция провела конференцию в Астане), *Regnum.Ru*, October 4, 2015, https://regnum.ru/news/polit/1984197.html.

Resalat Newspaper, Why Putin Travelled Iran? *Resalat Newspaper*, December 3, 2007, vista.ir/article/286484/چرا-پوتین-به-ایران-سفر-کرد؟.

Reuters, Army map suggests presence of Russian artillery unit in central Syria, November 18, 2015, www.reuters.com/article/us-mideast-crisis-syria-russia-military-idUSKCN0T72VW20151118.

Reuters, Around The World: Iranian Says Ties With Soviet Improve, *The New York Times*, February 10, 1986, www.nytimes.com/1986/02/10/world/around-the-world-iranian-says-ties-with-soviet-improve.html.

RFE, Presidents of Azerbaijan, Iran, And Russia Meet in Baku, *Radio Free Europe*, https://www.rferl.org/a/azerbaijan-iran-russia-presidents-meet-baku/27909041.html.

Richard Spencer, Vladimir Putin admits Bashar al-Assad responsible for Syrian uprising, *The Daily Telegraph*, June 11, 2013, www.telegraph.co.uk/news/worldnews/middleeast/syria/10114247/Vladimir-Putin-admits-Bashar-al-Assad-responsible-for-Syrian-uprising.html.

RNS, KAMAZ will supply to Iran in 2017 300 Trucks, December 14, 2016, https://rns.online/industry/KamaZ-postavit-v-Iran-v-2017-godu-300-mashinokomplektov-2016-12-14/.

Rouhollah.ir, The General Policies of the Islamic System During the Rebuilding of the Country, October 2, 1988, farsi.rouhollah.ir/library/sahifeh?volume=21&page=155.

RT, encourages Iran to join Russia-led Eurasian alliance, https://www.rt.com/politics/354743-putin-voices-support-to-iran/.

Russia interested in investment in Iran oil, gas projects, *Islamic Republic News Agency*, July 30, 2017, www8.irna.ir/en/News/82168725/.

Russia Iran relations and Pres. Rouhani's visit to Moscow, *Tehran Times*, April 8, 2017, https://www.pressreader.com/iran/tehran-times/20170408/281801398819035.

Russia proposes to Iran to organize licensed production of Russian T-90S tanks, February 03, 2016, tass.com/defense/854186.

Russia, China veto at U.N. on Syria chemical weapons is 'outrageous,' U.S. says, *The Washington Post*, February 28, 2017, https://www.washingtonpost.com/world/national-security/russia-china-veto-at-un-on-syria-chemical-weapons-is-outrageous-us-says/2017/02/28/c69adcf4-fdeb-11e6-99b4-9e613afeb09f_story.html?utm_term=.661523f59a8e.

Russian President Putin arrives in Tehran, www.presstv.ir/Detail/2015/11/23/438822/Russia-Putin-Tehran.

Security Council Unanimously Adopts Resolution 2254 (2015), Endorsing Road Map for Peace Process in Syria, Setting Timetable for Talks, www.un.org/press/en/2015/sc12171.doc.htm.

Security Council Unanimously Adopts Resolution 2254 (2015), Endorsing Road Map for Peace Process in Syria, Setting Timetable for Talks, SC/12171, Security Council 7588TH MEETING (PM), 18 December 2015.

Seyed Hossein Mousavian, The Iranian Nuclear Crisis: A Memoir (Washington: Carnegie Endowment for International Peace, 2012).

Shoayb Bahman, Suleimani-Puting Agreement (Tehran, Andisheh Sazan-e Noor) 2016.

Sputnik, Moscow supports Tehran's willing to become SCO joint member, Russian President Vladimir Putin said in an interview, https://sputniknews.com/world/201608051043962974-russia-iran-sco-putin/.

Sputnik, S-300 Was Just the Beginning: Iran Plans Huge Arms Deals With Russia, https://sputniknews.com/military/201602171034893494-russia-iran-military-deals/.

Sputnik, Syrian Opposition Signs Final Declaration After Marathon Talks in Astana, October 4, 2015, https://sputniknews.com/politics/201510041027986345-syrian-opposition-final-declaration/.

Sputnik, Vladimir Putin in the Land of Ayatollahs, *Sputnik*, November 27, 2015, https://sputniknews.com/radio_red_line/201511271030871131-vladimir-putin-in-the-land-of-ayatollahs/.

Su-30s, Mi-8s: Iran Considers Inking $8 Billion Arms Deal With Russia, https://sputniknews.com/military/201602161034839542-iran-russia-military-cooperation/.

TAAS, Iran plans negotiations with Russia on purchase of Sukhoi fighters, *TASS Russian News Agency*, August 22, 2016, tass.com/defense/895279.

TAAS, Iran's demand for Russian weaponry is estimated at $10 bln, *TASS Russian News Agency*, November 14, 2016, tass.com/defense/912141.

Telegraph, Roland Oliphant, Russia may sell Iran $10 billion worth of tanks and jets in new arms deal, *Telegraph*, November 14, 2016, www.telegraph.co.uk/news/2016/11/14/russia-may-sell-iran-10-billion-worth-of-tanks-and-jets-in-new-a/.

The formation of an organization of gas exporting countries, a guarantee for the future of energy trade, *Shana.ir*, April 9, 2007, www.shana.ir/fa/newsagency/101544/تشکیل-سازمان-کشورهای-صادر-کننده-گاز-تضمینی-برای-آینده-تجارت-انرژی.

The Guardian, More than 90%' of Russian airstrikes in Syria have not targeted Isis, U.S. says, *The Guardian*, October 7, 2015, https://www.theguardian.com/world/2015/oct/07/russia-airstrikes-syria-not-targetting-isis.

The Lethal Russian Air Power Deployed in Syria, *Forces Network News*, September 23, 2016, www.forces.net/news/tri-service/lethal-russian-air-power-deployed-syria.

The Syrian opposition signed the final declaration of the conference in Astana, *RIA Novosti*, October 4, 2015, https://ria.ru/world/20151004/1296432701.html.

Tyler, Rebuilding Plan to Cost $15 Billion, Tehran Says, *The Washington Post*, July 3, 1989.

UNP, Final Communiqué of the Action Group for Syria (Geneva Communiqué), June 30, 2012, peacemaker.un.org/node/1581.

USDOD, Department of Defense Press Briefing by Secretary Mattis, General Dunford and Special Envoy McGurk on the Campaign to Defeat ISIS in the Pentagon Press Briefing Room, https://www.defense.gov/News/Transcripts/Transcript-View/Article/1188225/department-of-defense-press-briefing-by-secretary-mattis-general-dunford-and-sp/.

Vestnik, Second round of Syrian opposition talks begins in Astana, October 2, 2015, vestnikkavkaza.net/news/Second-round-of-Syrian-opposition-talks-begins-in-Astana.html.

Vitaly Naumkin, Russia mulls strategy to solve Syrian conflict, June 3, 2015, www.al-monitor.com/pulse/originals/2015/06/russia-syria-conflict-strategy-moscow-meetings.html.

W&P, Russia will consider and "work out" possible applications of Iran (Россия рассмотрит и "будет прорабатывать" возможные заявки Ирана), *War and Peace*, December 25, 2007, www.warandpeace.ru/ru/news/view/18412/.

W&P, Russia will consider and "work out" possible applications of Iran (Россия рассмотрит и "будет прорабатывать" возможные заявки Ирана), *War and Peace*, December 25, 2007, www.warandpeace.ru/ru/news/view/18412/.

YJC, Suleimani-Putin Meeting in Moscow: Russia's Key to Attack Syria, *Yong Journalists Club*, www.yjc.ir/fa/news/5350142/ دیدار-سه-ساعته-پوتین-و-سردار-سلیمانی-در-مسکو-کلید-حمله-روسیه-به-سوریه.

CHAPTER 4

Iran and the European Union: Challenges and Opportunities

For most of its history, the EU has enjoyed robust relations with the Islamic Republic of Iran, an association bolstered by extensive trade ties. The ruinous war with Iraq made help from the EU even more urgent. By the time of the cease-fire in 1989, Iran's industry was operating at only 20–30 percent of capacity. Direct economic damage (i.e. physical destruction) inflicted on Iran amounted to a yearly average of 23.35 percent of the country's gross domestic product (GDP). The United Nations Security Council (UNSC) estimates that the cost of this damage, excluding military damage and loss of human life, amounted to $97.3 billion. By using a different exchange rate, the Iranian government puts the figure much higher, at over $300 billion. Military damage was estimated at $50 billion. Some observers estimate that the total cost of damage is close to $1000 billion.[1]

As president, Akbar Hashemi Rafsanjani pushed hard to improve the relations with the Europeans. With the United States out of the picture, the Iranian industry needed equipment, spare parts, and technological know-how, none more than the antiquated oil and gas infrastructure. Despite their concerns about the regime's revolutionary export and its

[1] Djamal Athari, 1991. "Revolutionary changes and post-war reconstruction in Iran." In: Reviving War Damaged Settlements. A Report and Charter prepared in connection with the Third International York Workshop on Settlement Reconstruction Post-War, 22–24 July 1991. York: University of York, Institute of Advanced Architectural Studies, Post-War Reconstruction and Development Unit, p. 14.

F. Rezaei, *Iran's Foreign Policy After the Nuclear Agreement*, Middle East Today, https://doi.org/10.1007/978-3-319-76789-5_4

destabilizing effect on the Middle East, the European countries, notably Germany, Great Britain, France, and Italy who depended on the Iranian oil, were ready to deal.[2]

To make the economic relations politically more palatable, the EU engaged in the so-called "critical dialogue," described as an attempt to change Iranian behavior and strengthen "moderate" forces in Iran. The EU was mostly interested in a dialogue on human rights, terrorism, and efforts to mediate Iran's implacable opposition to Israel. While critics described the "critical dialogue" as a fig leaf to legitimize trade with a rogue regime, the EU insisted that progress was made on these and other crucial issues.

This argument became untenable, however, when on September 17, 1992, Department 15 of the Ministry of Information, and Security (MOIS) assassinated several political dissidents in Europe. The high-profile murder of four Iranian Kurdish leaders—Mohammad Sadegh Sharafkandi, the exiled head of the Iranian Democratic Party of Kurdistan, Fattah Abdoli, Homayoun Ardalan and their translator Nouri Dehkordi—in the Mykonos restaurant in Berlin created a public firestorm. In 1997, a German court found that the secretive "Committee for Special Operations" which included the head of MOIS, Ali Fallahian, authorized the killings. The court alleged that Ayatollah Ali Khamenei, Hashemi Rafsanjani and the then foreign minister, Ali Velayati were privy to the decision.[3]

The ensuing diplomatic crisis put an end to the Critical Dialogue, but responding to the overtures of President Khatami's Dialogue Among Civilizations, the EU fashioned a new outreach known as Comprehensive Dialogue. Human rights and terrorism were still on the agenda, but the Comprehensive Dialogue negotiators hoped to sign a Trade and Cooperation agreement with the regime as a prelude to Iran's joining the World Trade Organization (WTO).

[2] Parvin Dadandish, Iran–Europe Relations: A Diagnostic Analysis, *Iranian Review of Foreign Affairs*, Vol. 3, No. 1, Spring 2012, pp. 57–88.

[3] M.V. Struwe, "The policy of "critical dialogue": an analysis of European human rights policy towards Iran from 1992 to 1997. Working Paper. University of Durham, Centre for Middle Eastern and Islamic Studies, Durham, 1998; Mary Williams Walsh, German Court Finds Iran's Leaders Ordered Slayings, *Los Angeles Times*, April 11, 1997, articles.latimes.com/1997-04-11/news/mn-47727_1_german-court; Roya Hakakian, The End of the Dispensable Iranian, *Spiegel*, April 10, 2007, www.spiegel.de/international/looking-back-at-the-mykonos-trial-the-end-of-the-dispensable-iranian-a-476369.html.

The EU and the Path Toward the JCPOA: From Mediator to a Key Player

Given these grandiose plans, the 2002 revelations of illicit nuclear activity in Iran came as a shattering disappointment. The disclosure, made by the Mujahedin-e Khalq Organization (MEK) and subsequently corroborated by the IAEA, came at a time of heightening attention to proliferation.

President George W. Bush had accused Saddam Hussein of harboring a WMD program and was preparing an attack on Iraq. While Great Britain participated in the 2003 invasion, other EU countries took a decidedly reserved stand. The subsequent chaos and destruction that the war had unleashed convinced many that nuclear diplomacy was preferable to a military conflict. It was against this background that the EU decided to launch a new and more robust Iran outreach.

The initiative was even more remarkable because it was the first effort of the EU to attempt "effective multilateralism" or "actorness". Coined by international relations experts, it was defined as an "effort to forge an effective posture as a multilateral actor." Tarja Cronberg, a member of the European Parliament in charge of the Iran portfolio, explained that nuclear multilateralism was a good fit for the "vision of the European Security Strategy and a founding general principle of EU foreign policy." Above all, the EU came to view itself as "a necessary counterbalance to U.S. militarism and Bush's unilateralism in Iraq." For her and other members of the European Union foreign policy team, "effective multilateralism was an explicit critique of the U.S. behavior in Iraq."[4]

In what was the first exercise in effective multilateralism, the EU challenged the unilateral sanctions that the United States imposed on Iran. The Iran–Libya Sanctions Act (ILSA) of 1996 authorized sanctions against foreign companies, most of them European, which had business ties to the oil and gas industry in Iran. These secondary sanctions angered the Europeans because the dominant position of the United States virtually forced these companies to choose the American market over Iran. The EU strongly protested the "extraterritoriality" of ILSA and even launched a complaint with the WTO.[5]

[4] Tarja Cronberg, Nuclear Unilateralism. Inside EU Negotiations (London: Routledge, 2017), loc. 220.

[5] Cornelius Adebahr, Europe and Iran: The Nuclear Deal and Beyond (London: Routledge, 2016).

President Mohammed Khatami who was looking for ways to break the nuclear impasse considered the EU initiative promising. On October 6, 2003, Khatami appointed Hassan Rouhani as the chief nuclear negotiator with the International Atomic Energy Agency (IAEA) in Vienna. Seyed Hossein Mousavian, a member of Rouhani's team, disclosed that at the time the Iranian leaders were engaged in a debate over the best-negotiating strategy. Some in the inner circle, including the former president Ayatollah Rafsanjani, wanted to deal directly with the United States. But the Supreme Leader Ayatollah Ali Khamenei strongly objected to the American option. Mousavian quoted Rafsanjani telling him that "in my time, the Americans showed signs of wanting to soften their stand," but we followed the policy of the leader [Khamenei] who did not favor normalization with the U.S." Mousavian noted that Khatami, who was also in favor of the American option, did not fare any better.[6]

Rouhani, who was an astute observer of the nuances of international politics, urged for the EU option. Echoing his boss, Mousavian explained the advantages of working with the Europeans and their newfound "effective multilateralism." Rouhani was convinced that the Iranians could split Brussels from Washington. As Mousavian put it, "interaction with the European Union could potentially deepen the transatlantic divide." Becoming more attuned to the EU vision, he added: "a European victory in resolving Iran's nuclear issue could establish the old continent as the champion of diplomatic and non-military approaches."[7]

Rouhani was even more enthusiastic about manipulating the relations between Washington and Brussels. In a remarkably candid memoir, he wrote that "from the beginning, the Americans repeatedly told the Europeans that the Iranians are insincere and misleading you, but the Europeans used to reply, 'we trust them'."[8] The disagreement between the Americans and the Europeans during the 2003 IAEA Board meeting was a case in point. While sharing the same goal of preventing Iran's proliferation, they differed on virtually everything else: referring the dossier to the United Nations Security Council (UNSC), imposing sanctions, and use of the military options. Mohammed ElBaradei, the IAEA chief, echoed Rouhani's observations that

[6] Seyed Hossein Mousavian, The Iranian Nuclear Crisis: A Memoir (Washington: Carnegie Endowment for International Peace, 2012), 82–83.

[7] Ibid., 87, 89.

[8] Hassan Rouhani, Amniat-e Meli va Diplomacy Hastei [National Security and Nuclear Diplomacy] (Tehran: Markaz-e Tahqiqat-e Istiratizhik, 2012).

there was an extraordinary rift among Western powers. While the United States, Canada, and Australia insisted on strong condemnation of Iran, Great Britain, France, and Germany tried to soften the resolution's tone.[9]

The quarrels among the allies provided Iran more breathing space and enabled it to advance its nuclear program. Overall, Iran's leaders felt that the strategy of "slow-plying" the international community through ongoing negotiations and simultaneously developing its nuclear program has served the country well. But Rouhani and Rafsanjani, who was still involved in the nuclear diplomacy, also understood that Tehran had to offer meaningful concessions. Under terms of the 2004 Paris Accord, Iran promised to temporarily suspend the production of uranium in exchange for lifting sanctions.[10]

Rouhani's strategy worked well until Mahmoud Ahmadinejad took office in 2005. The belligerent president who insisted on Iran's inalienable right to fabricate uranium had the IAEA seals removed soon after. Unsurprisingly, Ahmadinejad's hardball international politics, including threats to wipe out Israel and Holocaust denial, was not helpful to the EU's efforts to block American attempts to send the Iran dossier to the UNSC. Once the action switched to the United Nations, the Americans were better positioned to curtail the EU. As one expert explained, "once the U.N. Security Council was involved, the United States was in its established role," a development that affected the dynamics of the EU negotiators.[11]

Still, the EU was not about to abandon its first "multinationalism" project. Catherine Ashton, who took over from Javier Solana as the High Representative of the EU for Foreign Affairs and Security Policy on December 1, 2009, worked hard to restart the negotiations with Iran. However, four consecutive meetings—Istanbul in April 2012, Baghdad in May 2012, Moscow in June 2012, and Almaty in April 2013—failed to persuade the Principalists under Ahmadinejad to compromise.

By the time of his election to the presidency, Rouhani was firmly convinced that Iran would have to limit its nuclear program to normalize its international position. In 2015, after two years of arduous negotiations, Iran and the P5+1 (the UNSC five permanent members, China, France,

[9] Mohamed Elbaradei, The Age of Deception: Nuclear Diplomacy in Treacherous Times (New York: Metropolitan Books, 2011).

[10] Hassan Rouhani, Amniat-e Meli va Diplomacy Hastei [National Security and Nuclear Diplomacy] (Tehran: Markaz-e Tahqiqat-e Istiratizhik, 2012).

[11] Cornelius Adebahr, Europe and Iran: The Nuclear Deal and Beyond (London: Routledge, 2017), 56–67.

Russia, the United Kingdom, and the United States; plus Germany) reached the nuclear agreement formally known as the Joint Comprehensive Plan of Action (JCPOA). Although the Europeans played a secondary role in what became mostly bilateral negotiations between the United States and Iran, the EU considered the agreement to be a vindication of its signature multinationalism project. As Cronberg put it, "the EU is, as a result of the Iran deal, a prominent actor in non-proliferation…[which coordinated a major 12-year program]."[12]

Restarting the EU–Iran Relations: The Economic Dimension

Following the 1997 election of Reformist President Seyyed Mohammad Khatami, ties were greatly expanded in many areas, including trade, culture, and academic exchanges. EU exports to Iran increased from 3.9 billion euros ($4.3 billion) in 1996 to 11.3 billion euros ($12.6 billion) in 2006, while imports expanded from 5.8 billion euros ($6.4 billion) in 1996 to 14.1 billion euros ($15.7 billion) in 2006. In 2006, the EU was the destination of 38 percent of Iranian exports and accounted for 31 percent of Iranian imports.[13]

The volume of trade between Iran and the EU reached 23.5 billion euros in 2007, a decrease of 7 percent compared to 2006 (25.317 billion euros). According to the EU Statistics Office (Eurostat), of this amount, Iran's exports to the EU was 13.4 billion euros, a 3.6 percentage points decrease (14.126 billion euros) compared to 2006. Because of the 2007 global financial crisis, Iran's export to the EU decreased from 16 billion euros in 2008 to less than 10 billion euros in 2009. EU exports to Iran in 2007 totaled 10 billion euros, a 10.7 percent decrease in comparison to 2006 (11.191 billion euros). Thus, Iran had a trade surplus of 3.780 billion euros in its relations with the EU, an unprecedented amount since 2003. In the post-crisis years, trade relations between the two partners improved whereby in 2011 the value of goods exported from Iran to the EU exceeded 16 billion euros. But again, in 2012, because of sanctions, the value of Iran's export to the EU decreased to less than 6 billion euros, and EU exports to Iran plummeted to 7.3 billion euros.[14]

[12] Cronberg, Nuclear Unilateralism, loc. 205.

[13] EU and Iran Towards a New Partnership? *IRIS*, April 30, 2014, www.iris-france.org/44347-eu-and-iran-towards-a-new-partnership/.

[14] Ibid.

As the tables in Appendix I illustrate, in 2007 the EU exported to Iran 5.498 billion euros of machinery and vehicles, 2.356 billion euros of industrial products, 1.498 billion euros of chemicals, 166 million euros of food and agricultural products, and 242 million euros of raw materials. During the same period, Iran exported 12.158 billion euros of crude oil, 629 million euros industrial products, 420 million euros of chemical products, 275 million euros of food and agricultural products, 252 million euros of raw materials and 101 million euros of machinery. In 2007, Iran was identified as the fifth largest supplier of oil to the EU (After Russia, Norway, Libya, and Saudi Arabia) with exports of more than 12 billion euros, supplying 5.5 percent of the EU's demand for crude oil. However, Iran's crude exports to the EU in 2007 dropped by nearly 6 percent compared to the year 2006. The Iran's main trade partners among the EU members in 2007 were Italy, Germany, France, Spain, and the Netherlands.[15]

During the first half of 2015, Iran's trade with EU countries reached 5.4 billion euros, of which 4.6 billion euros comprised Iran's imports from the EU and 812 million euros of Iran's exports. During the same period, EU exports to Iran amounted to 5 percent more than the same period at 2014 and reached more than 3 billion euros. As a result, trade between Iran and the EU in the first six months of 2015 has grown by 7 percent. The value of bilateral trade in the first six months of 2014 was estimated at 3.4 billion euros. The European Union imported 550 million euros from Iran from January to June 2015, an increase of 14 percent compared to 480 million euros in the same period of 2014. The EU exports to Iran in the first half of 2015 grew by 5 percent, reaching 3.1 billion euros, compared to 2.9 billion euros in the same period in 2014. The trade balance of Iran's export–import in 2014 amounted to 5.240 billion euros in favor of the EU, compared to 4.662 billion in 2013.[16]

With the volume of trade already on the upswing, Iran was hoping for more valuable relations. With this mission in mind, President Rouhani left for Europe nine days after Implementation Day of the JCPOA on January 16, 2017. On his first stop in Rome, Prime Minister Mateo Renzi gave him a lavish welcome and Pope Francis invited him to the Vatican. Iran

[15] Economic relations of the Islamic Republic of Iran and the European Union, www.iranembassy.be/Farsi/HTM/ecoIranEU.htm.

[16] IBCCIM, From Diplomacy to Economic Relations, All you Need to Know about Relations Between Iran and European Countries, *Irano British Chamber of Commerce, Industries, and Mines*, www.ibchamber.org/azdiplomasitaravabeteeghtesadi/.

and Italy signed 14 billion euro ($18.4 billion) of cooperation agreements in the industrial fields including forestry, gas industry, road and railways buildings, mining and other infrastructure, banking, insurance, health, agriculture, shipbuilding, commerce, portal cooperation, medicine and pharmaceuticals, mostly with Italian companies Saipem, Fincantieri and Ansaldo Energy.[17]

ENI, the Italian oil company, signed a deal with Tehran to resume work in the oil and gas industry, including the third phase of the development of the Dakhokhin Oilfield and South Pars with a daily production capacity of 50 million cubic meters. The oil services group Saipem, a subsidiary of ENI, signed an agreement with Persian Oil & Gas Development Company to upgrade the Pars Shiraz and Tabriz refineries. Shana, the news website for the Iranian Oil Ministry, disclosed that both sides were mulling a pipeline contract worth at least $4.3 billion. Saipem was expected to control about 40 percent of any bilateral arrangement in the Iranian market.[18]

The Fincantieri shipbuilding company signed a framework accord with the Iranian Azim Gostaresh Hormoz Shipbuilding Industry Co. controlled by the large Iran Shipbuilding and Offshore Industries Complex Company. Fincantieri offered to construct merchant vessels and to supply Iran with locomotive engines. The Fincantieri subsidiary, Isotta Fraschini Motori also signed an agreement with Iran's Arka Tejarat Qeshm to set-up a joint proposal for the supply of 600 marine engines for smaller vessels. Another Fincantieri-connected company signed an agreement with Iran's holding group MAPNA for a possible joint proposal to supply the Iranian Railway with 70 shunting locomotive engines.[19]

In addition, Ansaldo Energia, an Italian engineering company, signed a deal with SUNIR Company (Iran's Power and Water Equipment and Service Export Company) according to which Ansaldo Energia agreed to

[17] IRNA, Iran, Italy sign 14 multi-billion euro cooperation agreements, *IRNA News Agency*, www.irna.ir/en/News/81935570; Shana, Details of Danieli's $5.7 Billion Investment in Iran, www.shana.ir/fa/newsagency/254271/جزئیات-سرمایه-گذاری-7-5-میلیارد-دلاری-دانیلی-ایتالیا-در-ایران; Taraz News, Iran–Italy billion-dollar contracts were signed, https://www.taraznews.com/content/135384.

[18] Iran Newspaper, Iran–Italy new oil talks start, *Iran Newspaper*, iran-newspaper.com/Newspaper/MobileBlock?NewspaperBlockID=114313; Shana, Iran, Italy Sign Up Giant Oil, Gas Agreement, 13 MoUs, *Shana.ir*, January 26, 2016, www.shana.ir/en/newsagency/254130/Iran-Italy-Sign-Up-Giant-Oil-Gas-Agreement-13-MoUs.

[19] Richard Tomkins, Fincantieri signs framework accords with Iranian companies, *UPI*, January 27, 2016, https://www.upi.com/Defense-News/2016/01/27/Fincantieri-signs-framework-accords-with-Iranian-companies/9091453927464/.

install new technology to minimize environmentally detrimental pollutants of energy exploration, such as gas flare ups.[20]

The Italian steelmaker Danieli group reached an agreement worth 5.7 billion euros to create a joint venture with Persian Metallic to deliver machinery and other equipment. In the infrastructure area, rail and highway projects, Iran signed a contract with Italian infrastructure company Condotte d'Acqua worth $4 billion. Italy's largest insurance company, Sacchet, has announced a $5 billion credit line for export to Iran.[21]

Ferrovie dello Stato (FS), Italy's state railway company signed a deal to supply assistance to the Iranian railways for both high-speed and conventional rail lines, including a complete program for training personnel. Under its 10-year plan, Iran was expected to increase its 15,000 km of rail tracks to 25,000 km by 2025. According to the semi-official Tasnim News Agency (TNA), Iran's Renewable Energy Organization also signed an agreement with Enel, an Italy-based multinational manufacturer and distributor of electricity and gas. According to the deal, Enel will invest in a geothermal-to-electrical power project in northwestern Iran.[22]

[20] Tehran Times, Italian firm offers eco-friendly technology to Iran, *Tehran Times*, November 2, 2016, www.tehrantimes.com/news/407948/Italian-firm-offers-eco-friendly-technology-to-Iran; Shana, Details of Danieli's $5.7 Billion Investment in Iran, www.shana.ir/fa/newsagency/254271/جزئیات-سرمایه-گذاری-5-7-میلیارد-دلاری-دانیلی-ایتالیا-در-ایران; Taraz News, Iran–Italy billion-dollar contracts were signed, https://www.taraznews.com/content/135384; Shana, Details of Danieli's $5.7 Billion Investment in Iran, www.shana.ir/fa/newsagency/254271/جزئیات-سرمایه-گذاری-5-7-میلیارد-دلاری-دانیلی-ایتالیا-در-ایران; Taraz News, Iran-Italy billion-dollar contracts were signed, https://www.taraznews.com/content/135384; ISNA, The necessity of joint investment in Iran and Italy, *Iranian Students News Agency*, www.isna.ir/news/94122314521/ضرورت-سرمایه-گذاری-مشترک-صنایع-ایران-و-ایتالیا.

[21] Tehran Times, Italian firm offers eco-friendly technology to Iran, *Tehran Times*, November 2, 2016, www.tehrantimes.com/news/407948/Italian-firm-offers-eco-friendly-technology-to-Iran; Shana, Details of Danieli's $5.7 Billion Investment in Iran, www.shana.ir/fa/newsagency/254271/جزئیات-سرمایه-گذاری-5-7-میلیارد-دلاری-دانیلی-ایتالیا-در-ایران; Taraz News, Iran-Italy billion-dollar contracts were signed, https://www.taraznews.com/content/135384; Shana, Details of Danieli's $5.7 Billion Investment in Iran, www.shana.ir/fa/newsagency/254271/جزئیات-سرمایه-گذاری-5-7-میلیارد-دلاری-دانیلی-ایتالیا-در-ایران; Taraz News, Iran-Italy billion-dollar contracts were signed, https://www.taraznews.com/content/135384; ISNA, The necessity of joint investment in Iran and Italy, *Iranian Students News Agency*, www.isna.ir/news/94122314521/ضرورت-سرمایه-گذاری-مشترک-صنایع-ایران-و-ایتالیا.

[22] Shana, Details of Danieli's $5.7 Billion Investment in Iran, www.shana.ir/fa/newsagency/254271/جزئیات-سرمایه-گذاری-5-7-میلیارد-دلاری-دانیلی-ایتالیا-در-ایران; Taraz News, Iran–Italy billion-dollar contracts were signed, https://www.taraznews.com/content/135384; PressTV, Italy, Iran to sign €1.2bn rail deal next week, *PressTV*, July 5, 2017, www.presstv.ir/Detail/2017/07/05/527485/Iran-Italy-rail-deal-transport-FS-Mazzoncini;

During the second leg of the trip, on January 27, Rouhani traveled to Paris and met the French president, François Hollande. Rouhani and his delegation signed score of contracts worth around 40 billion euros including the purchase of 118 Airbus airplanes estimated at $27 billion. The contract comprised 46 of the narrow-body A320 family including the A321 model, 38 long-haul A330s, 12 double-decker 380 super-cargo aircraft and 16 of Europe's newest long-range model, the A350. According to the deal, Airbus agreed to provide the Iranians with pilot training and maintenance services.[23]

The French oil company Total signed a deal for the purchase of up to 200,000 barrels of oil from Iran, especially for refineries in France and Europe. According to the deal, Total also agreed to examine the potential of the Iranian oil industry to assess its wider presence. A joint venture between French car manufacturer Peugeot Citroen and the Iranian carmaker Iran-Khodro renewed an old collaboration. The French company had broken off relations in 2012 after the French firm went into partnership with General Motors in the United States. According to the TNA, Iran-Khodro finalized the new deal worth 500 million euros with Peugeot in which the latter agreed to produce the Peugeot 2008 crossover, 208 supermini, and 301 compact cars under the 50–50 joint venture. The French company will also produce 200,000 vehicles a year using parts fabricated inside Iran.[24]

Italy's Enel seeks investment in Iran energy, *Trend*, January 2016, https://en.trend.az/business/energy/2485323.html.

[23] Iran's $51 billion agreement with Italy and France, *Middle East Bank*, January 30, 2016, middleeastbank.ir/page/94–11-10; The contract for the purchase of 118 Airbus was signed, https://www.tasnimnews.com/fa/news/1394/11/09/984424/قرارداد-خرید-118-ایرباس-امضا-شد-جزئیات-کامل-و-جدول.

[24] Charles Kennedy, Iran Signs Oil Deal with Total, Deal Done In Euros, *Oil Price*, February 08, 2016, oilprice.com/Energy/Energy-General/Iran-Signs-Oil-Deal-With-Total-Deal-Done-In-Euros.html; TNA, Primary 500-Million-euro-Agreement between Iran Khodro and France Peugeot, *Tasnim News Agency*, January 24, 2016, https://www.tasnimnews.com/fa/news/1394/11/04/979789/توافق-اولیه500-میلیون-یورویی-ایران-خودرو-با-پژوی-فرانسه; Dolate Bahar, Agreement with Peugeot is signed, *Dolate Bahar*, www.dolatebahar.com/view/213672/قرارداد-بازگشت-«پژو»-به-بازار-ایران-روز-پنجشنبه-امضا-می‌شود; Radio Farda, Signing a contract between Saipa and Citroën, https://www.radiofarda.com/a/f3-citreon-saipa-business-iran/27879379.html; The contract for the purchase of 118 Airbus was signed, https://www.tasnimnews.com/fa/news/1394/11/09/984424/قرارداد-خرید-118-ایرباس-امضا-شد-جزئیات-کامل-و-جدول; IRNA, Total Returns to Iran, www.irna.ir/fa/News/82585512.

Alstom, a French multinational company engaged in rail transport engineering, joined Iran's Industrial Development and Renovation Organization (IDRO), and Iranian Rail Industries Development Co. (IRICO) to modernize rail services. Accordingly, Alstom will invest 60 percent of the project, while each of the Iranian companies will invest 20 percent. The Alstom initiative complemented a July 2015 $7 million deal by the architectural firm AREP to redevelop three key railway stations in Iran.[25]

French CMA CGM, the third-largest container line, and the Iranian state-owned carrier signed a preliminary agreement outlining how the duo plans to cooperate. CMA CGM agreed to open six offices in Iran—in Tehran, Bandar Abbas, Bushehr, Khorramshahr, Bandar Imam Khomeini and Chabahar—effective April 1, 2017, to become a "key player" in Iran's global trade.[26] Iran's National Water and Wastewater Engineering Company (NWWEC) has also signed an agreement with French company Suez aimed at refurbishing and maintaining NWWEC treatment facilities. Sanofi, one of the largest French pharmaceutical companies also signed a memorandum of co-operation with the Iran Food and Drug Administration by which Sanofi will use its expertise to improve the healthcare landscape in the middle-eastern country.[27]

On July 3, 2017, Iran and French Total signed a $4.8 billion agreement, which would lead to technology transfer to Iran. By signing this agreement to develop the 11th phase of the South Pars, 56 million cubic meters of gas per day will be added to the capacity of Iran to extract from the joint gas field with Qatar. According to experts, the development of this phase is equivalent to developing an oil field with a daily capacity of 410 thousand barrels per day.[28]

[25] IFP, Iran, France's Alstom Sign Deal for Building Railroad Cars, *Iran-France Page*, July 25, 2017, ifpnews.com/wired/iran-frances-alstom-sign-deal-building-railroad-cars/.

[26] Reuters, CMA CGM Steps Up Iran Presence with IRISL Deal, *Reuters*, January 28, 2016, https://gcaptain.com/cma-cgm-steps-up-iran-presence-with-irisl-deal/?utm_source=feedburner&utm_medium=twitter&utm_campaign=Feed%3A+Gcaptain+%28gCaptain.com%29#.VqpyrVI0uf8; Lemonde, CMA CGM Inks Deal with IRISL to Increase Presence in Iran, *World Maritime News*, January 29, 2017, worldmaritimenews.com/archives/181826/cma-cgm-inks-deal-with-irisl-to-increase-presence-in-iran/.

[27] Pharmfile, Sanofi expands Iranian presence with memorandum agreement, January 29, 2016, www.pharmafile.com/news/502619/sanofi-expands-iranian-presence-memorandum-agreement.

[28] Euronews, The $4.8 Billion Development of South Pars is Signed with Total, *Euronews*, July 3, 2017, fa.euronews.com/2017/07/03/iran-total-contract-push.

On June 22, 2017, two Iranian privately owned carriers, Zagros Airlines and Iran Airtour, signed a memorandum of understanding to buy new Airbus aircrafts. Zagros Airlines opted for 28 new planes, while Iran Airtour decided on 48 aircrafts. In total, Rouhani's deals with Italy and France reached $51.4 billion.[29]

Iran pursued smaller but important commercial ties with other EU countries as well. For instance, the Iran Investment Co-operation Commission (IICC) established the Center for Iran and Europe Trade Development in Prague. According to Reza Alikhanzadeh, a representative of Iran Cooperative Chamber, the IICC key objective is to identify European companies and encourage them to invest in Iran. Alikhanzadeh argued that the reason for choosing the Czech Republic is its advanced industries, which has led many German companies to operate in this country, and in terms of Czech's location, it could be a gateway for Iran's entry into Europe.[30]

On June 1, 2016, Iran's Ministry of Industry, Mines, and Trade established the Iranian-European Economic Cooperation Center (IEECC) in Germany to renew Iran–EU relations. The center offered comprehensive industrial, mineral, and commercial banks services as well as services related to investment in Iran.[31]

Many of the new transactions were made possible when the Belgian-based Society for Worldwide Interbank Financial Telecommunication (SWIFT) reconnected the Central Bank of Iran (CBI) and 15 non-sanctioned Iranian banks in February 15, 2016 to its service, (the banks having been disconnected in March 2012), allowing them to resume cross-border payment transfers related to oil shipments and other trade with foreign counterparts. In addition, international branches of Iranian banks opened in Europe and were reconnected to Trans-European Automated Real-time Gross

[29] Airbus, Zagros Airlines places a commitment for 28 new Airbus aircraft, *Airbus*, June 22, 2017, www.airbus.com/newsroom/press-releases/en/2017/06/zagros-airlines-places-a-commitment-for-28-new-airbus-aircraft-.html; Airbus, Iran Airtour commits to 45 A320neo aircraft, *Airbus*, June 22, 2017, www.airbus.com/newsroom/press-releases/en/2017/06/iran-airtour-commits-to-45-a320neo-aircraft.html; Iran's $51 billion agreement with Italy and France, *Middle East Bank*, January 30, 2016, middleeastbank.ir/page/94–11-10; Details of Rouhani's travel deals to Italy and France, www.asriran.com/fa/news/447823/جزئیات-قراردادهای-سفر-روحانی-به-ایتالیا-و-فرانسه.

[30] Asr Bazar, Establishment of Iran and Europe Trade Development Center, *Asr Bazar*, www.asretebar.com/News/93861/ایجاد-مرکز-توسعه-تجارتایران-و-اروپا.

[31] Donyaye Eghtesad, Iran-European Business-Economic Cooperation Center opened, June 1, 2016, donya-e-eqtesad.com/news/1054660.

Settlement Express Transfer System (TARGET 2), a leading European payment platform owned and operated by the Eurosystem for processing large-value payments.[32]

The International Monetary Fund (IMF) estimated that the volume of foreign investment in Iran would reach $6 billion by 2022, the end of Rouhani presidency. Should this figure materialize, it would, according to the IMF's report, have an unprecedented effect in Iran's annual growth.[33]

Impressive as this progress was, the total volume of trade was less than at the pre-2002 level. The Iranian authorities blamed the Trump administration and the Republican-dominated Congress for many of their problems. But lacking leverage in the United States, the Rouhani government hoped to mobilize the EU to plead its case.

The EU to the Defense: Fighting the American Sanctions

With the prospect of lucrative economic ties, the EU has emerged as a strong supporter of President Rouhani's efforts to rejoin the community of nations. Protecting the integrity of the JCPOA has been a top priority in this context.

But developments in Washington created difficulties for the EU. On December 1, 2016, the U.S. Senate passed a 10-year extension of sanctions against Iran known as the Iran Sanctions Act (ISA). Iran's foreign minister Mohammad Javad Zarif appealed to the EU, which convened a high-level international commission in Vienna on January 10, 2017. Abbas Araqchi, Iran's deputy foreign minister and Helga Schmid EU's political director co-chaired the meeting. The Iranian officials have railed against ISA, denouncing it as a violation of the accord. "We will not tolerate that a state party to the JCPOA violates its commitments under various pretexts. The Islamic Republic of Iran will provide appropriate response to such moves," Araqchi told the Islamic Republic News Agency (IRNA) in

[32] Reuters, Iranian banks reconnected to SWIFT network after four-year hiatus, *Reuters*, February 17, 2016, https://www.reuters.com/article/us-iran-banks-swift/iranian-banks-reconnected-to-swift-network-after-four-year-hiatus-idUSKCN0VQ1FD.

[33] IMF, Islamic republic of Iran selected issues, *International Monetary Fund*, February 2017, Report No. 17/63; OIETAI, The IMF's latest Estimate of the Growth of Foreign Investment in Iran in the 12th Government, *Organization for Investment Economic and Technical Assistance of Iran*, www.investiniran.ir/fa/news/5050/ جدیدترین-برآورد-صندوق-بین-المللی-پول-از-روند-رشد-سرمایه-گذاری-خارجی-در-ایران-در-دولت-دوازدهم.

a subsequent interview.[34] All along, EU officials had warned the Trump administration not to destroy the nuclear deal with an abundance of new sanctions. An EU delegation had also met with senior Republicans in Congress to deliver a message that new economic sanctions on Iran are "a non-starter."[35]

Despite the hard lobbying against the sanctions, the EU registered little if any success because of a high-profile effort by the anti-JCPOA lobby. A carryover from the opposition to the original nuclear deal, the pro-Israel lobby vowed to undermine the agreement. United Against Nuclear Iran (UANI), a flagship project of the Foundation for the Defense of Democracies (FDD), described itself as a "private sanctions campaigns and state and Federal legislative initiatives focus on ending the economic and financial support of the Iranian regime by corporations, firms, entities and individuals at a time when the international community is attempting to compel Iran to abandon its illegal nuclear weapons program, support for terrorism and gross human rights violations."[36]

The Republican-dominated Congress moved to impose a long series of non-nuclear related sanctions. Introduced in the Senate on March 23, Congress voted 98–92 for new sanctions on Iran known as "Countering Iran's Destabilizing Activities Act of 2017" (CIDAA-2017), which directs the Departments of State, Treasury and Defense, and the Director of National Intelligence to submit a strategy every two year for deterring conventional and asymmetric Iranian activities that threaten the United States and its key allies in the Middle East, North Africa, and beyond.[37]

The CIDAA-2017 imposed new restrictions on Iran's ballistic missile program, and restrictions on arms sales, and also penalized Iran for its alleged sponsorship of terrorism and human rights violations. The bill proposed to freeze assets and sanctioned any individual that contributes to

[34] Reuters, Extension of Iran Sanctions Act passes U.S. Congress, December 1, 2016, www.reuters.com/article/us-iran-nuclear-usa-sanctions-idUSKBN13Q5JW; EUEA, 25 April 2017 meeting of the JCPOA Joint Commission, *European Union External Action*, April 25, 2017, https://eeas.europa.eu/headquarters/headquarters-homepage_en/25069/25%20April%202017%20meeting%20of%20the%20JCPOA%20Joint%20Commission; Financial Tribune, JCPOA Panel Meets on Tehran Grievances, *Financial Tribune*, January 11, 2017, https://financialtribune.com/articles/national/57225/jcpoa-panel-meets-on-tehran-grievances.

[35] John Dyer, "EU warns Trump not to Destroy Nuclear Deal," *Financial Times*, December 21, 2016.

[36] Iran Business Risk, https://www.unitedagainstnucleariran.com/.

[37] Congress.gov, S.722 – Countering Iran's Destabilizing Activities Act of 2017, June 29, 2017, https://www.congress.gov/bill/115th-congress/senate-bill/722.

Iran's ballistic missile and nuclear programs, or the sale or transfer of specified military equipment or the provision of related technical or financial assistance to Iran. Furthermore, the bill blocked the Revolutionary Guards' properties and sanctioned its affiliated foreign individuals, and prohibits transactions with any Iranian individual who supports or has ties to terrorism. In other words, the CIDAA-2017 sanctions the entire Revolutionary Guards organization as a terror group. Moreover, the CIDAA-2017 sanctions legislation imposes asset blocking measures against any individual responsible for killings, torture, or other forms of human rights violations in Iran.[38]

The anti-Iran measures followed the February 2017 ban on banking transfers by the U.S. Department of the Treasury's Office of Foreign Assets Control (OFAC) that was imposed on 25 Iranian individuals and companies, among them the Iran-based company Matin Sanat Nik Andishan, and three associated Chinese companies which supported or contributed to Iran's ballistic missile program by supplying missile-related items. The OFAC also sanctioned two senior Iranian defense officials, Morteza Farasatpour, Director of Iran's Shahid Bagheri Industries Group (SBIG), and Rahim Ahmadi, coordinator of missile flight tests with Iran's Aerospace Industries Organization (AIO), for their involvement in Iran's ballistic missile program.[39]

Acting as a private sanction manager, UANI has identified a long list of companies to which it sends an open letter to dissuade them from doing business with Iran: "They should reconsider their interest, given the continuing legal, financial, and reputational risks of doing business in Iran. Given the significant risks that remain, responsible companies should exercise caution and continue to forgo Iran business until Iran demonstrates sustained compliance with the JCPOA and fundamentally abandons its pursuit of nuclear weapons, sponsorship of terrorism, human rights violations, and destabilizing aggression in the Middle East."[40]

Under the energetic leadership of Federica Mogherini who replaced Catharine Ashton as the High Representative, the EU made its opposition to altering the terms of the JCPOA well known. Mogherini and other EU

[38] Ibid.

[39] USDOT, Treasury Sanctions Iranian Defense Officials and a China-Based Network for Supporting Iran's Ballistic Missile Program, U.S. Department of Treasury, May 17, 2017, https://www.treasury.gov/press-center/press-releases/Pages/sm0088.aspx.

[40] Ibid.

officials have stressed that the JCPOA is an international agreement which cannot be abrogated by unilateral actions of one of its co-signers. In a statement, Mogherini wrote that "this was an historic achievement for the security of the security of the region and the whole world, a success for multilateral diplomacy that has proven to work and deliver." She added that the "deal belongs to the international community, having been endorsed by the U.N.S.C., that expects all sides to keep the commitments they took two years ago." On this and other occasions, Mogherini made sure to emphasize that Russia and China, the other signatories of the deal, were fully supportive of the EU position. Not coincidentally, the EU echoed closely the position of the Rouhani government on the binding character of the JCPOA.[41]

Even though the Trump administration recertified Iran as compliant with the JCPOA in April and July 2017, American non-nuclear sanctions have taken a toll on the economic normalization process. On July 12, 2017, the National Security, and Foreign Policy Committee of the Majlis published a report accusing the U.S. of intimidating major foreign financial institutions from doing business with Iran. The report noted that "even though the JCPOA clearly calls for sanctions relief, unfortunately, the P5+1 countries (the US, Russia, China, France, Britain, and Germany) seem not willing to implement the given promises in the agreement." The Committee members explained that the Americans used a flimsy pretext for their decision. The Committee added, "we should ask how can small-size European banks or other countries do business with Iran but not the large banks? Aren't they [small and large banks] operating based on the same system and regulations?" The report concludes: "It seems that Iran sanctions have been lifted on paper only and it has to do with the threats emanating from the United States to the European banks. They are afraid of facing heavy penalties in dealing with Iran."[42]

Government officials echoed this complaint. Valliolah Seif, the Governor of the Central Bank, stated that "the reason that European banks are hesitating or maybe they don't have the courage to work with Iranian banks, it's because of the heavy penalties that they have been going through or impose upon them [by Americans]. And also, they have been asked not to work with Iranian banks. And they are afraid. If these threats are removed,

[41] Statement of the Second Anniversary of the JCPOA, *European Union External Action*, July 17, 2017.

[42] FNA, Full Text of Majlis Report Regarding JCPOA Implementation, *Fars News Agency*, www.farsnews.com/newstext.php?nn=13960421001267.

Iranian banks can reestablish their connections with international banks."[43] Parviz Aghili, head of the private Middle East Bank in Tehran noted that "for the moment, the little European banks have agreed to work with us. They include banks from Italy, Austria, Switzerland, Germany, and Belgium. But not a single medium-sized or big bank has so far agreed to do it."[44]

Javad Zarif asserted that the U.S. hasn't been holding up its part of the bargain to normalize economic relations. "I believe the United States is showing minimal compliance with the JCPOA but certainly the United States, particularly the new administration has not taken the necessary measures in order to be in full compliance. It would take some political will on the part of the United States to live up its commitment under the deal. I believe it is important for everybody to try to implement it in good faith and I believe at the end of the day it will be in the interest of everybody that the Iranian population would see the economic benefits of the deal."[45]

During the fifth Joint Commission of the JCPOA meeting held in the Austrian capital on April 25, 2017, Iran denounced the U.S. threats to renege on the JCPOA and quoted from Trump to that effect. Iran has filed an official complaint to the Joint Commission against the new U.S. sanctions. Iranian Majlis Speaker Ali Larijani alleged that the U.S. has violated the letter of the JCPOA. "Iran will determinedly prevent the U.S. from reaching its objectives. Diplomatic actions should be taken regarding the new U.S. sanctions and a complaint should be lodged to the related committee which has been done. The committee has issued an announcement according to which the U.S. has committed some violations in implementing the JCPOA."[46]

[43] FNA, The Major Reasons for Foreign Banks not Working with us, *Fars News Agency*, www.farsnews.com/13951110000333.

[44] Asharq Alawsat, Foreign Banks Reluctant to Deal with Iran, *Asharq Alawsat*, September 14, 2016, https://english.aawsat.com/theaawsat/business/foreign-banks-reluctant-deal-iran.

[45] Nilo Tabrizy, Iran's Top Diplomat Discusses Economic Impact of Nuclear Deal, *The New York Times*, July 24, 2017, https://www.nytimes.com/video/world/middleeast/100000005289800/irans-nuclear-deal-mohammad-zarif.html.

[46] TNA, U.S. Lack of Commitment to JCPOA Raised in Joint Commission Meeting: Iran, *Tasnim News Agency*, May 01, 2017, https://www.tasnimnews.com/en/news/2017/05/01/1394430/us-lack-of-commitment-to-jcpoa-raised-in-joint-commission-meeting-iran; TT, Iran says has filed complaint to JCPOA Joint Commission over U.S. sanctions, *Tehran Times*, August 1, 2017, www.tehrantimes.com/news/415606/Iran-says-has-filed-complaint-to-JCPOA-Joint-Commission-over.

Mogherini and other EU officials explained that the old and new American sanctions created the phenomenon of "overcompliance." In Cronberg's view, "the remaining U.S. [nonnuclear] sanctions, possible penalties, and the threat of being closed off from the U.S. financial markets creates fear. This fear lingers in European banks today." Cronberg listed "complex U.S. legislative system dealing with money laundering, dollar transitions, terrorism financing, and the like as well as risk related to possible new sanctions." Several European bankers explained how overcompliance was built into their investment decisions.[47] A French banker noted that "We have abolished certain operations because we fear expensive fines similar to these imposed on BNP Paribas, Crédit Agricole, and Societe Generale Bank Jordan." These banks were fined for violation of U.S. economic sanctions imposed against Iran. A German banker added: "The choice is easy between huge interests in the U.S. compared to complicated and risky ones in Iran."[48]

To contravene the American pressure, the EU has sought to plug Iran into the international economic system. Helping to put the Iranian banking system in order was among the most important steps. On June 12, 2016, Iran joined the Eurasian Group Combating Money Laundering and Financing of Terrorism, a regional body to assist it with Anti-Money Laundering (AML) and Counter Terrorism Financing (CFT) protocols. Western countries, including Britain, offered to assist three Iranian banks—Persia International Bank, Bank Melli Iran (BMI), and Sepah Bank—to upgrade their policies.[49]

The BMI and Sepah banks, once the premier outlets for illicit transactions, announced that they would no longer service accounts connected with the Revolutionary Guards, including that of the giant construction company Khatam al-Anbiya. Other major Iranian financial institutions, such as Bank Keshavarzi (Agribank) and Bank Maskan (Housing Bank), ceased to handle the Guards' business as well. Rouhani was rewarded

[47] Cronberg, Nuclear Unilateralism, loc. 1958; Parvaz Ahgaili, "Foreign Banks Reluctant to Deal with Iran," *Asharq al Awsat*, September 14, 2016.

[48] Mutlaq Muneer, U.S. Bans European Banks From Funding Investment in Iran, *Asharq Alawsat*, March 3, 2017, https://english.aawsat.com/mutlaq-muneer/business/us-bans-european-banks-funding-investment-iran.

[49] FNA, Eurasian Group Combating Money Laundering and Financing of Terrorism, June 12, 2016, *Fars News Agency*, www.farsnews.com/13950323000767; FT, British regulators help Iranian banks come in from the cold, *Financial Times*, January 31, 2016, https://www.ft.com/content/54144ec2-c692-11e5-b3b1-7b2481276e45.

when, on June 23, 2017, the plenary session of the Financial Action Task Force (FATF) decided to continue its suspension of the countermeasures against Iranian banks—a step demanded by the European Union to help Iranian banks to develop ties with larger European banks. This step was essential because small banks can extend letters of credit of only \$10–20 million.[50]

In yet another effort to normalize Iran's standing, the EU has renewed its efforts to convince the WTO to accept Tehran. The EU had worked on this issue before 2002, but abandoned the plan after the MEK revealed Iran's illicit nuclear sites in Natanz and Fordow. In his visit to Tehran, Federica Mogherini, the EU foreign policy chief announced that EU will support Iran to join the WTO. However, because of American resistance, Iran's membership quest was stalled.[51]

The EU, fully aware of the complexities of the negotiated political order, was eager to help the normalizers against the challenges of the hardliners. Because of the wave of new sanctions, the latter portrayed Rouhani as a naïve loser who had given up the nuclear project for little gain. Indeed, the Islamic Revolutionary Guards Corps (IRGC) and their supporters claimed that Iran was the victim of a "grand conspiracy to deny Iran access to the international banking network." The hardline Principalists turned out in numbers to protest against Rouhani and even heckled him during the Quds Day parade in June. Observers speculated that such an unprecedented breach of public protocol was coordinated by some of the vigilante groups associated with the Basij.[52]

More astounding, the Haqqani School clergy and other hard core Principalists, with the tacit support of the Supreme Leader, have tried to

[50] Mashregh, Self-Sanctioning of Revolutionary Organizations, *Mashregh News*, https://www.mashreghnews.ir/news/626806/واکنش-بانک-های-ملتو-سپه-به-خودتحریمی-نهادهای-انقلابی-سازندگی; FT, FATF Extends Suspension of Iran Countermeasures – Update, *Financial Tribune*, June 25, 2017, https://financialtribune.com/articles/economy-business-and-markets/67137/fatf-extends-suspension-of-iran-countermeasures-update.

[51] EU Supports Iran Joining WTO, *Euronews*, April 16, 2016, fa.euronews.com/2016/04/16/eu-and-iran-turn-a-new-page-in-diplomatic-relations; TNA, EU announced its Support of Iran's Entering WTO, *Tasnim News Agency*, https://www.tasnimnews.com/fa/news/1395/05/10/1144674/حمایت-اتحادیه-اروپا-از-الحاق-ایران-به-سازمان-تجارت-جهانی.

[52] Mashregh News, Attacking the President will be in whose Interest, *Mashregh News*, June 23, 2017, https://www.mashreghnews.ir/news/741126/به-آب-جمهور-رئیس-علیه-شعار-ریزد-می-آسیاب-کدام; Farhad Rezaei, Iran Complains U.S. Pressure on Foreign Banks Violates JCPOA, Atlantic Council, July 25, 2017, www.atlanticcouncil.org/blogs/iraninsight/iran-complains-us-pressure-on-foreign-banks-violates-jcpoa.

minimize Iran's engagement with the world community. *Sobh-e-No*, a newspaper close to the Principalists, objected to Rouhani's agreement to purchase Boeing planes. Under the headline "A deal to create 100,000 jobs in U.S.," the paper wrote that the deal "will support tens of thousands of U.S. jobs directly associated with production and delivery of the 777-300ERs and nearly 100,000 jobs in the U.S. aerospace value stream."[53] On January 30, 2017, the semi-official Fars News Agency (FNA) quoted Brigadier General Mohammad Reza Naghdi, the commander of the Basij and the cultural deputy of the IRGC. According to the FNA, Naghdi stated that "they [the Rouhani administration] had promised to force the United States to return the $2 billion they owe us [as part of Iran's frozen assets in the United States], but now they have poured another $16 billion in their [Americans'] mouths."[54] Naghdi slammed the president for purchasing planes from Western aircraft manufacturers: "The occurrences of the Rome and France trips [of Rouhani] were a stab to the body of the Resistance Economy. These measures are against the Resistance Economy, and in case of continuance of such measures, we should close down the technical universities of the country."[55]

While the hardliners failed to stop President Rouhani from engaging with the EU on economic issues, they used the human rights issue to hamper his progress.

EU's Human Rights Convention and Iran: A Discordant Tone in a Warm Relationship

From its very inception, the EU showed great concern for human rights. In 1950, the then European Council drafted the Convention for the Protection of Human Rights and Fundamental Freedoms. Subsequently renamed the European Convention on Human Rights (ECHR), it was ratified in September 1953. Initially binding only on EU members, the ECHR standards have been gradually expanded to support human rights on a global

[53] Sobhe Sadegh No, A Deal to Create 100,000 Jobs in America, sobhe-no.ir/newspaper/141/3/4677; Boeing, Boeing, Iran Air Announce Agreement for 80 Airplanes, boeing.mediaroom.com/2016-12-11-Boeing-Iran-Air-Announce-Agreement-for-80-Airplanes.

[54] FNA, Relations with America has Nothing Except Loss, *Fars News Agency*, www.farsnews.com/newstext.php?nn=13950922001227.

[55] Naghdi: the Rome and France trips were a Stab to the Body of the Resistance Economy, *Dana.ir*, www.dana.ir/news/629320.html/فتونیوز-رخدادهای-سفر-رم-و-فرانسه-خنجری-بر-پیکر-اقتصاد-مقاومتی-است.

scale. As of 1990, the European Parliament has awarded the annual Sakharov Prize for Freedom of Thought, better known as the Sakharov Prize, to individuals and groups who "dedicated their lives to the defense of human rights and freedom of thought."[56]

As noted above, human rights were part of "dialogue" with Iran, an engagement that the EU tried to bolster over the years. Reflecting these concerns, in 2012, the European Parliament awarded the Sakharov Prize to Jafar Panahi and Nasrin Sotoudeh, a film director and a human rights activist. When the two were imprisoned by the Iranian authorities, the EU human rights push was compromised. Cronberg recalled that as the chair of the European Parliament subcommittee on Iran, she was scheduled to travel to Iran in 2011 with a delegation of parliamentarians. After being refused a visit to the jailed Panahi and Sotoudeh, the trip was delayed until Rouhani's presidency made it possible in 2014. Cronberg's human rights activism did not go unnoticed by the hardliners. In 2014, the Basij awarded her the Wet Gunpower Prize which, in the words of the official description, "ridiculous people whose chaffy character is evident to everyone and when this nature of theirs is accompanied with self-belief turns into an indefinite foolishness for them."[57]

The EU hoped that following the JCPOA, the Rouhani government would be in a better position to improve the human rights record. But, as Chap. 1 indicated, parts of the law enforcement community such as MOIS, the IRGC-Basij, and the judiciary are resistant to state supervision. Defeated on the nuclear deal, the hardliners retaliated against Rouhani by ramping up human rights violations, as Chap. 2 described. Unsurprisingly, the Principalists blocked the opening of an EU embassy in Tehran after the European Parliament voted to normalize relations with Iran on October 25, 2016. While the government welcomed the decision, the chief of the human rights council in the judiciary Mohammed-Javad Larijani stated that "after establishing the office, they want to establish close contacts with human rights advocates so they can ensure the process of human rights implementation. Thus, they should know that the judicial system will definitely not allow such a nest of corruption to be established in Iran." The Mizan news agency associated with the judiciary run a story under the headline "The European Parliament's issuing a document for Iran's slavery."

[56]The European Parliament Supports Human Rights, http://www.europarl.europa.eu/sakharovprize/en/sakharov-prize-.html.

[57]Cronberg, Nuclear Unilateralism, loc. 78.

The article stated: "The EU Parliament resolution provides grounds for cooperation with Iran, but it urges that the cooperation be based on talks that should push Iran toward a more appropriate behavior in the region, observing human rights and economic transparency. In reality, the conditions that Europe has set for normalization of ties with Iran are exactly the [same] policies and demands that the United States has [for Iran]. This cooperation document, which in reality is a document of slavery, shows that Europe has still not separated its way from the United States." Ayatollah Sadeq Amoli Larijani, the chief of the judiciary, dismissed the allegation of human rights abuses in Iran and accused the EU of "making instrumental use" of the issue. In his words, "human rights bodies have turned into an instrument to put independent countries under pressure."[58]

Faced with a defiant hardline position on human rights, the EU was forced to extend its human rights sanctions on Iran. The European Council announced in an April 11, 2017 statement that the sanctions would be extended until April 2018 in response "to serious human rights violations in Iran." The restrictive measures included a travel ban and an asset freeze against 82 people and one entity as well as a ban on exports to Iran of equipment for monitoring telecommunications and other gear that "might be used for internal repression."[59]

Among the listed was Saeed Mortazavi, former prosecutor general of Tehran; Maj-Gen Seyyed Hassan Firuzabadi, former chief of staff of Iran's Armed Forces; Mohammad Moghisseh (a.k.a. Nasserian), head of Tehran Revolutionary Court, branch 28; and Abbas Pir-Abbassi, magistrate of a criminal chamber. IRGC Brigadier General Mohammad Hejazi, Abdolghassem Salavati, head of Tehran Revolutionary Court,

[58] FNA, Supreme Leader Urges Judiciary to Take Action against Sponsors of Iran Sanctions, *Fars News Agency*, July 3, 2017, en.farsnews.com/newstext.aspx?nn=13960412001404; Rohallah Faghihi, "Will Iran Allow the EU to Open Shop in Tehran? *Al Monitor*, October 27, 2016; PressTV, "EU Extends Sanctions on Iran over Human Rights Claims," *PressTV*, April 11, 2017.

[59] Iran: Council extends by one year sanctions responding to serious human rights violation, *European Council of the European Union*, April 11, 2017, www.consilium.europa.eu/en/press/press-releases/2017/04/11-iran-sanctions/: Regulation (EU) 2017/77 of 16 January 2017 implementing Regulation 267/2012 and Decision 2017/83 of 16 January 2017 amending Decision 2010/413/CFSP.; Implementing Regulation 2017/685 of 11 April 2017 implementing Regulation (EU) No 359/2011 and Council Decision (CFSP) 2017/689 of 11 April 2017 amending Decision 2011/235/CFSP. http://eur-lex.europa.eu/legal-content/EN/TXT/?uri=uriserv:OJ.L_.2017.099.01.0010.01.ENG&toc=OJ:L:2017:099:TOC.

branch 15; and Mahmoud Abbaszadeh-Meshkini, secretary of the Human Rights Council were also featured. Interestingly enough, the EU added Behrouz Kamalian, head of the IRGC-linked "Ashiyaneh" cyber group. PressTV, which reported on the EU extension of the human right censure, repeated judiciary chief Ayatollah Sadeq Amoli Larijani's contention that the West has made "instrument use" of the issue."[60]

While the extension might have satisfied the EU's human rights concerns, the punishment was largely symbolic, at least in comparison with the long list of human rights violations published by Human Rights Watch in its Iran 2016 report. As mentioned in the previous chapters, in 2015 the number of executions reached 977 people, compared to 743 the year before. Other forms of persecution and harassment have also increased. The statistics for the first half of 2016 seem to be even worse. On July 29, the Iran human rights organization watchdogs reported 386 executions in 2016.[61]

Well-informed regime insiders considered the EU report a largely symbolic reprimand, which was not likely to affect economic relations. Iranian and international human rights activists came to the same conclusion when Mogherini did not respond to their urgent appeals to crack down on Iran. Ironically, the feeble EU response bolstered the confidence of the hardliners, leading to more human rights infractions.

Much as the human rights issue embarrassed the moderates, it was not their most serious hurdle on the road to normalization. The IRGC-QF increased boldness in exporting the revolution was, by far, a more grievous problem.

[60] Implementing Regulation 2017/685 of 11 April 2017 implementing Regulation (EU) No 359/2011 and Council Decision (CFSP) 2017/689 of 11 April 2017 amending Decision 2011/235/CFSP, http://eur-lex.europa.eu/legal-content/EN/TXT/?uri=uriserv:OJ.L_.2017.099.01.0010.01.ENG&toc=OJ:L:2017:099:TOC.

[61] PressTV, EU Extends Sanctions on Iran over Human Rights Claims," *PressTV*, April 11, 2017; Amnesty International, Death Penalty 2015: Alarming Surge In Recorded Executions Sees Highest Toll In More Than 25 Years, *Amnesty International*, April 6, 2016, https://www.amnesty.org/en/press-releases/2016/04/alarming-surge-in-recorded-executions-sees-highest-toll-in-more-than-25-years/; IHRDC, IHRDC Chart of Executions by the Islamic Republic of Iran – 2016, *Iran Human Rights Documentation Center*, July 29, 2017, www.iranhrdc.org/english/publications/human-rights-data/chart-of-executions/1000000620-ihrdc-chart-of-executions-by-the-islamic-republic-of-iran-2016.html.

Conclusion

Iran and the EU have a mostly cooperative history. Relations became complicated when the EU tried to persuade Iran to roll back the nuclear program. When these efforts failed, the EU joined in the international sanctions. The sanction regime proved to be economically costly for the EU because Iran has been a hugely important market for EU goods, services, and an investment destination.

Since the JCPOA was implemented in January 2016, Iran boosted a range of EU industries and services, including a large contract for the Airbus. Both the normalizers and the Principalists agree that the European connection is important to the future of Iran, even if the pivot to the East is successful.

From the perspective of Tehran, the EU serves as a buffer against the Trump administration's efforts to weaken the JCPOA. When Trump indicated his plan to abrogate the deal, the EU protested loudly. Federica Mogherini took steps to close loopholes and clarify some oversight provisions to prevent the United States from charging Iran with noncompliance.

The moderates and the hardliners harbor a different view of several issues that undergird the relations with the EU. The degree of collaboration with the EU is the factor in this context. The EU would like to develop a true economic, mutually beneficial partisanship, while Iran seems to opt for a narrow economic collaboration to preserve the so-called resistance economy.

The competing elites differ on how to respond to the political demands of the EU on the issue of human rights, curbing terrorism, and dealing with Syria. As a rule, the EU leaders would like Iran to improve their human rights record, rein in the terrorism that destabilizes the region and work to remove Bashar al-Assad. While the normalizers would like to show some flexibility, the hardliners are not likely to give in on any of these issues.

References

Airbus, Iran Airtour commits to 45 A320neo aircraft, *Airbus*, June 22, 2017, www.airbus.com/newsroom/press-releases/en/2017/06/iran-airtour-commits-to-45-a320neo-aircraft.html.

Airbus, Zagros Airlines places a commitment for 28 new Airbus aircraft, *Airbus*, June 22, 2017, www.airbus.com/newsroom/press-releases/en/2017/06/zagros-airlines-places-a-commitment-for-28-new-airbus-aircraft-.html.

Amnesty International, Death Penalty 2015: Alarming Surge In Recorded Executions Sees Highest Toll In More Than 25 Years, *Amnesty International*, April 6, 2016, https://www.amnesty.org/en/press-releases/2016/04/alarming-surge-in-recorded-executions-sees-highest-toll-in-more-than-25-years/.

Asharq Alawsat, Foreign Banks Reluctant to Deal with Iran, *Asharq Alawsat*, September 14, 2016, https://english.aawsat.com/theaawsat/business/foreign-banks-reluctant-deal-iran.

Asr Bazar, Establishment of Iran and Europe Trade Development Center, *Asr Bazar*, www.asretebar.com/News/93861/ایجاد-مرکز-توسعه-تجارت-ایران-و-اروپا.

AsrIran, Details of Rouhani's travel deals to Italy and France, *Asr Iran*, www.asriran.com/fa/news/447823/جزئیات-قراردادهای-سفر-روحانی-به-ایتالیا-و-فرانسه.

Boeing, Boeing, Iran Air Announce Agreement for 80 Airplanes, boeing.mediaroom.com/2016-12-11-Boeing-Iran-Air-Announce-Agreement-for-80-Airplanes.

Charles Kennedy, Iran Signs Oil Deal with Total, Deal Done In Euros, *Oil Price*, February 08, 2016, oilprice.com/Energy/Energy-General/Iran-Signs-Oil-Deal-With-Total-Deal-Done-In-Euros.html.

Congress.gov, S.722 – Countering Iran's Destabilizing Activities Act of 2017, June 29, 2017, https://www.congress.gov/bill/115th-congress/senate-bill/722.

Cornelius Adebahr, Europe and Iran: The Nuclear Deal and Beyond (London: Routledge, 2016).

Djamal Athari, 1991. "Revolutionary changes and post-war reconstruction in Iran." In: Reviving War Damaged Settlements. A Report and Charter prepared in connection with the Third International York Workshop on Settlement Reconstruction Post-War, 22–24 July 1991. York: University of York, Institute of Advanced Architectural Studies, Post-War Reconstruction and Development Unit.

Dolate Bahar, Agreement with Peugeot is signed, *Dolate Bahar*, www.dolatebahar.com/view/213672/قرارداد-بازگشت-«پژو»-به-بازار-ایران-روز-پنجشنبه-امضا-می‌شود.

Donyaye Eghtesad, Iran-European Business-Economic Cooperation Center opened, June 1, 2016, donya-e-eqtesad.com/news/1054660.

Economic relations of the Islamic Republic of Iran and the European Union, www.iranembassy.be/Farsi/HTM/ecoIranEU.htm.

EU and Iran Towards a New Partnership?, *IRIS*, April 30, 2014, www.iris-france.org/44347-eu-and-iran-towards-a-new-partnership/.

EU Supports Iran Joining WTO, *Euronews*, April 16, 2016, fa.euronews.com/2016/04/16/eu-and-iran-turn-a-new-page-in-diplomatic-relations.

EUEA, 25 April 2017 meeting of the JCPOA Joint Commission, *European Union External Action*, April 25, 2017, https://eeas.europa.eu/headquarters/headquarters-homepage_en/25069/25%20April%202017%20meeting%20of%20the%20JCPOA%20Joint%20Commission.

Euronews, The $4.8 Billion Development of South Pars is Signed with Total, *Euronews*, July 3, 2017, fa.euronews.com/2017/07/03/iran-total-contract-push.

Farhad Rezaei, Iran Complains U.S. Pressure on Foreign Banks Violates JCPOA, *Atlantic Council*, July 25, 2017, www.atlanticcouncil.org/blogs/iraninsight/iran-complains-us-pressure-on-foreign-banks-violates-jcpoa.

Financial Tribune, JCPOA Panel Meets on Tehran Grievances, *Financial Tribune*, January 11, 2017, https://financialtribune.com/articles/national/57225/jcpoa-panel-meets-on-tehran-grievances.

FNA, Eurasian Group Combating Money Laundering and Financing of Terrorism, June 12, 2016, *Fars News Agency*, www.farsnews.com/13950323000767.

FNA, Full Text of Majlis Report Regarding JCPOA Implementation, *Fars News Agency*, www.farsnews.com/newstext.php?nn=13960421001267.

FNA, Relations with America has Nothing Except Loss, *Fars News Agency*, www.farsnews.com/newstext.php?nn=13950922001227.

FNA, Supreme Leader Urges Judiciary to Take Action against Sponsors of Iran Sanctions, *Fars News Agency*, July 3, 2017, en.farsnews.com/newstext.aspx?nn=13960412001404.

FNA, The Major Reasons for Foreign Banks not Working with us, *Fars News Agency*, www.farsnews.com/13951110000333.

FT, British regulators help Iranian banks come in from the cold, *Financial Times*, January 31, 2016, https://www.ft.com/content/54144ec2-c692-11e5-b3b1-7b2481276e45.

FT, FATF Extends Suspension of Iran Countermeasures – Update, *Financial Tribune*, June 25, 2017, https://financialtribune.com/articles/economy-business-and-markets/67137/fatf-extends-suspension-of-iran-countermeasures-update.

Hassan Rouhani, Amniat-e Meli va Diplomacy Hastei [National Security and Nuclear Diplomacy]. (Tehran: Markaz-e Tahqiqat-e Istiratizhik, 2012).

IBCCIM, From Diplomacy to Economic Relations, All you Need to Know about Relations Between Iran and European Countries, *Irano British Chamber of Commerce, Industries, and Mines*, www.ibchamber.org/azdiplomasitarava beteeghtesadi/.

IFP, Iran, France's Alstom Sign Deal for Building Railroad Cars, *Iran-France Page*, July 25, 2017, ifpnews.com/wired/iran-frances-alstom-sign-deal-building-railroad-cars/.

IHRDC, IHRDC Chart of Executions by the Islamic Republic of Iran – 2016, *Iran Human Rights Documentation Center*, July 29, 2017, www.iranhrdc.org/english/publications/human-rights-data/chart-of-executions/1000000620-ihrdc-chart-of-executions-by-the-islamic-republic-of-iran-2016.html.

IMF, Islamic republic of Iran selected issues, *International Monetary Fund*, February 2017, Report No. 17/63.

Implementing Regulation 2017/685 of 11 April 2017 implementing Regulation (EU) No 359/2011 and Council Decision (CFSP) 2017/689 of 11 April 2017 amending Decision 2011/235/CFSP, http://eur-lex.europa.eu/legal-content/EN/TXT/?uri=uriserv:OJ.L_.2017.099.01.0010.01.ENG&toc=OJ:L:2017:099:TOC.

Iran Business Risk, https://www.unitedagainstnucleariran.com/.

Iran Newspaper, Iran–Italy new oil talks start, *Iran Newspaper*, iran-newspaper.com/Newspaper/MobileBlock?NewspaperBlockID=114313.

Iran: Council extends by one year sanctions responding to serious human rights violation, *European Council of the European Union*, April 11, 2017, www.consilium.europa.eu/en/press/press-releases/2017/04/11-iran-sanctions/.

Iran's $51 billion agreement with Italy and France, *Middle East Bank*, January 30, 2016, middleeastbank.ir/page/94-11-10.

IRNA, Iran, Italy sign 14 multi-billion euro cooperation agreements, *IRNA News Agency*, www.irna.ir/en/News/81935570.

IRNA, Total Returns to Iran, www.irna.ir/fa/News/82585512.

ISNA, The necessity of joint investment in Iran and Italy, *Iranian Students News Agency*, www.isna.ir/news/94122314521/ضرورت-سرمایه-گذاری-مشترک-صنایع-ایران-و-ایتالیا.

Italy's Enel seeks investment in Iran energ, *Trend*, January 2016, https://en.trend.az/business/energy/2485323.html.

John Dyer, "EU warns Trump not to Destroy Nuclear Deal," *Financial Times*, December 21, 2016.

Lemonde, CMA CGM Inks Deal with IRISL to Increase Presence in Iran, *World Maritime News*, January 29, 2017, worldmaritimenews.com/archives/181826/cma-cgm-inks-deal-with-irisl-to-increase-presence-in-iran/.

M.V. Struwe, "The policy of "critical dialogue": an analysis of European human rights policy towards Iran from 1992 to 1997. Working Paper. University of Durham, Centre for Middle Eastern and Islamic Studies, Durham, 1998.

Mary Williams Walsh, German Court Finds Iran's Leaders Ordered Slayings, *Los Angeles Times*, April 11, 1997, articles.latimes.com/1997-04-11/news/mn-47727_1_german-court.

Mashregh News, Attacking the President will be in whose Interest, *Mashregh News*, June 23, 2017, https://www.mashreghnews.ir/news/741126/شعار-علیه-رئیس-جمهور-آب-به-کدام-آسیاب-می-ریزد.

Mashregh News, Self-Sanctioning of Revolutionary Organizations, *Mashregh News*, https://www.mashreghnews.ir/news/626806/واکنش-بانک-های-ملت-و-سپه-به-خودتحریمی-نهادهای-انقلابی-سازندگی.

Mohamed Elbaradei, The Age of Deception: Nuclear Diplomacy in Treacherous Times (New York: Metropolitan Books, 2011).

Mutlaq Muneer, U.S. Bans European Banks From Funding Investment in Iran, *Asharq Alawsat*, March 3, 2017, https://english.aawsat.com/mutlaq-muneer/business/us-bans-european-banks-funding-investment-iran.

Naghdi: the Rome and France trips were a Stab to the Body of the Resistance Economy, *Dana.ir*, www.dana.ir/news/629320.html/فتونیوز-رخدادهای-سفر-رم-و-فرانسه-خنجری-بر-پیکر-اقتصاد-مقاومتی-است.

Nilo Tabrizy, Iran's Top Diplomat Discusses Economic Impact of Nuclear Deal, *The New York Times*, July 24, 2017, https://www.nytimes.com/video/world/middleeast/100000005289800/irans-nuclear-deal-mohammad-zarif.html.

OIETAI, The IMF's latest Estimate of the Growth of Foreign Investment in Iran in the 12th Government, *Organization for Investment Economic and Technical Assistance of Iran*, www.investiniran.ir/fa/news/5050/جدیدترین-برآورد-صندوق-بین-المللی-پول-از-روند-رشد-سرمایه-گذاری-خارجی-در-ایران-در-دولت-دوازدهم.

Parvaz Ahgaili, "Foreign Banks Reluctant to Deal with Iran," *Asharq al Awsat*, September 14, 2016.

Parvin Dadandish, Iran–Europe Relations:A Diagnostic Analysis, *Iranian Review of Foreign Affairs*, Vol. 3, No. 1, Spring 2012.

Pharmfile, Sanofi expands Iranian presence with memorandum agreement, January 29, 2016, www.pharmafile.com/news/502619/sanofi-expands-iranian-presence-memorandum-agreement.

PressTV, Italy, Iran to sign €1.2bn rail deal next week, *PressTV*, July 5, 2017, www.presstv.ir/Detail/2017/07/05/527485/Iran-Italy-rail-deal-transport-FS-Mazzoncini.

PressTV, "EU Extends Sanctions on Iran over Human Rights Claims," *PressTV*, April 11, 2017.

Radio Farda, Signing a contract between Saipa and Citroën, https://www.radiofarda.com/a/f3-citreon-saipa-business-iran/27879379.html.

Regulation (EU) 2017/77 of 16 January 2017 implementing Regulation 267/2012 and Decision 2017/83 of 16 January 2017 amending Decision 2010/413/CFSP.

Reuters, CMA CGM Steps Up Iran Presence with IRISL Deal, *Reuters*, January 28, 2016, https://gcaptain.com/cma-cgm-steps-up-iran-presence-with-irisl-deal/?utm_source=feedburner&utm_medium=twitter&utm_campaign=Feed%3A+Gcaptain+%28gCaptain.com%29#.VqpyrVI0uf8.

Reuters, Extension of Iran Sanctions Act passes U.S. Congress, December 1, 2016, www.reuters.com/article/us-iran-nuclear-usa-sanctions-idUSKBN13Q5JW.

Reuters, Iranian banks reconnected to SWIFT network after four-year hiatus, *Reuters*, February 17, 2016, https://www.reuters.com/article/us-iran-banks-swift/iranian-banks-reconnected-to-swift-network-after-four-year-hiatus-idUSKCN0VQ1FD.

Richard Tomkins, Fincantieri signs framework accords with Iranian companies, *UPI*, January 27, 2016, https://www.upi.com/Defense-News/2016/01/27/Fincantieri-signs-framework-accords-with-Iranian-companies/9091453927464/.

Rohallah Faghihi, "Will Iran Allow the EU to Open Shop in Tehran? *Al Monitor*, October 27, 2016.

Roya Hakakian, The End of the Dispensable Iranian, *Spiegel*, April 10, 2007, www.spiegel.de/international/looking-back-at-the-mykonos-trial-the-end-of-the-dispensable-iranian-a-476369.html.

Seyed Hossein Mousavian, The Iranian Nuclear Crisis: A Memoir (Washington: Carnegie Endowment for International Peace, 2012).

Shana, Details of Danieli's $5.7 Billion Investment in Iran, www.shana.ir/fa/newsagency/254271/جزئیات-سرمایه-گذاری-5-7-میلیارد-دلاری-دانیلی-ایتالیا-در-ایران.

Shana, Iran, Italy Sign Up Giant Oil, Gas Agreement, 13 MoUs, *Shana.ir*, January 26, 2016, www.shana.ir/en/newsagency/254130/Iran-Italy-Sign-Up-Giant-Oil-Gas-Agreement-13-MoUs.

Sobhe Sadegh No, A Deal to Creat 100,000 Jobs in America, sobhe-no.ir/newspaper/141/3/4677.

Statement of the Second Anniversary of the JCPOA, *European Union External Action*, July 17, 2017.

Taraz News, Iran-Italy billion-dollar contracts were signed, https://www.taraz-news.com/content/135384.

Tarja Cronberg, Nuclear Unilateralism. Inside EU Negotiations (London: Routledge, 2017), loc. 220.

Tehran Times, Italian firm offers eco-friendly technology to Iran, *Tehran Times*, November 2, 2016, www.tehrantimes.com/news/407948/Italian-firm-offers-eco-friendly-technology-to-Iran.

The contract for the purchase of 118 Airbus was signed, https://www.tasnimnews.com/fa/news/1394/11/09/984424/قرارداد-خرید-118-ایرباس-امضا-شد-جزئیات-کامل-و-جدول.

The European Parliament Supports Human Rights, http://www.europarl.europa.eu/sakharovprize/en/sakharov-prize-.html.

TNA, EU announced its Support of Iran's Entering WTO, *Tasnim News Agency*, https://www.tasnimnews.com/fa/news/1395/05/10/1144674/حمایت-اتحادیه-اروپا-از-الحالق-ایران-به-سازمان-تجارت-جهانی.

TNA, Primary 500-Million-euro-Agreement between Iran Khodro and France Peugeot, *Tasnim News Agency*, January 24, 2016, https://www.tasnimnews.com/fa/news/1394/11/04/979789/توافق-اولیه500-میلیون-یورویی-ایران-خودرو-با-پژوی-فرانسه.

TNA, U.S. Lack of Commitment to JCPOA Raised in Joint Commission Meeting: Iran, *Tasnim News Agency*, May 01, 2017, https://www.tasnim-news.com/en/news/2017/05/01/1394430/us-lack-of-commitment-to-jcpoa-raised-in-joint-commission-meeting-iran.

TT, Iran says has filed complaint to JCPOA Joint Commission over U.S. sanctions, *Tehran Times*, August 1, 2017, www.tehrantimes.com/news/415606/Iran-says-has-filed-complaint-to-JCPOA-Joint-Commission-over.

USDOT, Treasury Sanctions Iranian Defense Officials and a China-Based Network for Supporting Iran's Ballistic Missile Program, U.S. Department of Treasury, May 17, 2017, https://www.treasury.gov/press-center/press-releases/Pages/sm0088.aspx.

CHAPTER 5

Iran and Iraq: The Lebanonization Project in the Balance

For Ayatollah Ruhollah Khomeini, Iraq represented a perfect terrain for spreading the Islamic revolution. The Iraqi Shiites, estimated at some 60 percent of the population, fit all the criteria of the mustazafeen, the "oppressed." Considerably poorer than their Sunni compatriots, they had no official role in the Baathist regime and were subject to brutal oppression. Members of the Dawa Party, a significant underground movement, were persecuted and many killed, exiled or escaped to Iran. Indeed, the Supreme Leader hoped to replace the Baathist regime with the Islamic Republic of Iraq. He expected that there would be no shortage of Iraqi Shiites who were willing to play a role in this venture.

On September 15, 1981, Ayatollah Khamenei, the then president, accompanied by a high ranking Revolutionary Guard delegation traveled to Syria where a group of prominent Shiite leaders lived. He visited Ayatollah Mohammad Baqir al-Hakim, Ayatollah Mohammad Mahdi Asefi (who later became speaker of the Dawa Party) and Ali al-Kourani, a senior Lebanese cleric who studied in Najaf and later moved to Iran. In 1982, these Shiite leaders established the Supreme Council for the Islamic Revolution in Iraq (SCIRI), the Iraqi Shiite Islamist Iraqi political party, and Ayatollah Mahmoud Hashemi Shahroudi, member of Islamic Dawa Party, led the council.[1]

[1] Travel to Syria, *Khamenei.ir*, September 6, 1984, farsi.khamenei.ir/print-content?id=8699.

F. Rezaei, *Iran's Foreign Policy After the Nuclear Agreement*, Middle East Today, https://doi.org/10.1007/978-3-319-76789-5_5

Working closely with the Revolutionary Guards commander, Major General Ismail Daghayeghi, SCIRI adopted a two-pronged plan to fight the Baathists. The organization established secret resistance cells in Iraq and mobilized expatriate Iraqis to train them as an armed force. Abdel Aziz al-Hakim, Ayatollah al-Hakim's brother, became the head of its military unit, the Badr Brigade (later Badr Corps), which fought with the Islamic Revolutionary Guards Corps (IRGC) during the Iran–Iraq war.[2]

The Ramadan base of the IRGC-QF located in Kermanshah coordinated guerrilla operations of Iraqi mujahedin groups during the Iran–Iraq war. One of them was Harakat ul-Mujahidin, and other was the veteran Munazzamat al-'Amal al-Islami (Islamic Action Call). In 1984, the Badr Brigade joint the military operations led by several noted commanders. Hadi Farhan Abdullah Al-Ameri, known as Abu Hassan Al-Ameri, Abu Ali al-Basri, Abu Lagha, and Jamal Jaafar Ibrahimi a.k.a. Abu Mahdi al-Muhandis, commanded the organization. Major General Mohammad Ali Rahmani who led the mission, pointed out that "the IRGC supported these groups to carry out jihadist operations inside Iraqi territory." But Iran's failure to win the war ended Khomeini's vision for an Islamic Republic of Iraq.[3]

Iraq's invasion of Kuwait in 1990 and the subsequent American-led Gulf War of 1991 seemed to offer a new opportunity for revolutionary export. Acting on the command from SCIRI, the secret cells, aided by the Badr fighters, and operatives from the QF launched a rebellion against the regime. Even though weakened by the war, Saddam Hussein managed to put the uprising down in his customary brutal fashion, an action which left thousands of Shiites dead, wounded, imprisoned, or displaced. In Tehran, the lesson of the failed rebellion was sober. As one observer put it, "as long as he [Saddam Hussein] remained [in power]…[there could be no] maturation of the Islamic revolution, which they believed would restore dignity to the Islamic world."[4]

Indirectly, 9/11 turned out to be the answer that the revolutionary exporters were waiting for. While Iraq was not involved in the al-Qaeda terror attack, neoconservatives in the administration of George W. Bush

[2] Global Security, *Badr Corps*, www.globalsecurity.org/military/world/para/badr.htm.

[3] Defa, Formation of Badr Army, the Factor for Successful Operations Against Baath Regime, defapress.ir/fa/news/26472/تشکیل-سپاه-بدر-عامل-یکپارچگی-عملیات-نظامی-علیه-رژیم-بعث.

[4] Afshon Ostovar, Vanguard of the Imam: Religion, Politics, and Iran's Revolutionary Guards (London: Oxford University Press, 2016), 1878.

sought ways of getting rid of Saddam Hussein whom they were suspecting of harboring a nuclear program. For some of the more ambitious neoconservatives, however, regime change in Iraq was just part of a larger plan to undermine Iran's bourgeoning hold on the Middle East. In 1999, David Wurmser, then a fellow at the American Enterprise Institute (AEI), the flagship of the neoconservative movement, wrote a book, *Tyranny's Ally*, advocating for this plan. Wurmser noted that the "overthrow of Saddam Hussein" would "sever Iran from its Shiite foundation [in Iraq]." The Iraqi Shiites, if liberated from tyranny, could be expected to present a challenge to Iran's influence and revolution. He explained that the "spiritual faith characteristic of these Muslims is seldom compatible with zealotry." In other words, "the good Shiites of Iraq" would vanquish the "bad Shiites" of Iran.[5]

While Wurmser, the Middle East adviser to Vice President Dick Cheney, had little understanding of the reality in Iraq, he could count on Ahmad Chalabi to promote the war against Saddam Hussein. Chalabi, the head of the Iraqi National Congress (INC), had extensive ties to the IRGC-QF and persuaded the Vice President that SCIRI would support the American invasion. In August 2002, Chalabi brought Ayatollah Abdel-Aziz al-Hakim to a meeting with the Secretary of Defense Donald Rumsfeld in which Cheney participated via videoconferencing. Al-Hakim gave SCIRI's blessing to the plans to remove Saddam Hussein without mentioning its deep ties with the Iranian regime.[6]

It was only after the invasion in March 2003 that the Americans learned of the extent of the Iranian involvement in Iraq.

Lebanonization of Iraq: Laying the Foundations of the Project

Even for Ayatollah Ali Khamenei, a man who viewed history through a spiritual perspective, the removal of Saddam Hussein was nothing short of a miracle. Speaking at a Friday prayers on April 11, 2003, Khamenei said "Saddam is gone, and the Iranian nation is happy for this. It's been twenty years that the Iranian nation says 'death to Saddam.' Now death came to

[5] David Wurmser, Tyranny's Ally (Washington, DC: American Enterprise Institute, 1999); quoted in Aram Rostam, 163.

[6] Aram Rostam, The Man Who Pushed America to War: The Extraordinary Life, Adventures, and Obsessions of Ahmad Chalabi, 223; I Worship the God of Ayatollah Sistani, *Bashgah Andishe*, January 27, 2004, www.bashgah.net/fa/content/show/5637.

Saddam. The happiness of our nation in this matter is like the happiness of Iraqi people. Our stance is just like the Iraqi nation. Iraqi people are happy with Saddam's demise. We're also happy. Saddam was a dictator, a bad guy, a cruel, dishonest and promise breaker, and very evil … He was very bad for both the Iraqi people and us as a neighbor."[7]

Relying less on the divine, the IRGC-QF moved swiftly to expand its influence. Qassem Suleimani, the then little-known QF commander in charge of the complex operation, would later reveal his strategy. He noted that the Revolutionary Guards aimed at adding Iraq to its "Shiite empire" regional project, and undermining and demoralizing the United States. Drawing heavily on the deceptive practices first developed in Lebanon, the IRGC-QF used a mix of hard and soft power delivered through its extensive network of proxies. In the words of one analyst, "From Day 1, Iran saw its opportunity: a chance to make a client state of Iraq. In that contest, Iran won, and the United States lost."[8]

Even before the U.S. forces arrived in Baghdad, on December 22, 2002, the Supreme National Security Council (SNSC) chaired by the Ayatollah Ali Khamenei discussed "a possible U.S. attack on Iraq," The SNSC decided to provide the Quds Force with $26 million to rebuild and strengthen Iraqi Badr Brigade. Elements of the IRGC-QF together with hundreds of Badr Corp fighters crossed the porous border of southern Iraq, where they joined up with the cells around Basra. They set up several camps on the Iran–Iraq border for smuggling weapons into Iraq.[9]

But Suleimani did not have an entirely free hand since the moderate government of President Mohamad Khatami urged restraint, and the Supreme Leader was not sure whether they planned to invade Iran a well, a rumor which circulated in Tehran in the early months of 2003. The Ministry of Intelligence and Security (MOIS), and the Intelligence Division of IRGC, which watched the widespread looting, argued that the Americans would be easy to unsettle because they had neither sufficient manpower nor the political will to police Iraq. Overruling Khatami, the IRGC-QF decided to unleash the anti-American insurgency.

[7] Speech in Friday Prayers, *Khamenei.ir*, April 11, 2003, farsi.khamenei.ir/speech-content?id=3168.

[8] Afshon Ostovar, Vanguard of the Imam: Religion, Politics, and Iran's Revolutionary Guards (London: Oxford University Press, 2016), 5067; Tim Arango, "U.S. Sees a Vital Toll Road, but Iran Sees a Threat," *New York Times*, May 27, 2017.

[9] Joseph Felter and Brian Fishman, Iranian Strategy in Iraq: Politics and "Other Means" (West Point, NY: Combating Terrorism Center, West Point, 2008), pp. 17, 29; Summer Basri, "Iran- Iraq: Can We Handle a Menace?" *NewsBlaze*, April 25, 2009, 9.

Initially, the QF worked with the Badr Brigade, but it soon reached out to several Shiite insurgent groups. The Mahdi Army, founded by Muqtada al-Sadr, son of the revered Ayatollah Mohammed Sadiq al-Sadr and the son-in-law of the Grand Ayatollah Mohammed Baqir al-Sadr, became an early partner. The QF could also count on the Asa'ib Ahl al-Haq (League of the Righteous, also known as the Khazali Network) and Khataib Hezbollah (Battalions of Hezbollah), the al-Shabani network and TharAllah (Vengeance of God). But when some of the groups proved unmanageable, the QF used its old Lebanese technique of manipulation and splintering to reconfigure the insurgency. In the end, Suleimani settled on three groups: The Promised Day Brigade (PDB), the successor of the Mahdi Army, the Asa'ib Ahl al-Haq and the Kataib Hezbollah to wage a bloody war against the Americans and unleash a sectarian mayhem.

By 2004, the bloodshed was difficult to ignore. The IRGC supplied the Shiite militias with an array of weapons such as rockets, magnetic bombs, and Improvised Rocket-Assisted Mortars (IRAMs). Improvised Explosive Devices (IEDs) and the more lethal Iranian-made Explosively Formed Penetrators (EFPs) were used as road bombs. Militants were sent to special camps in Iran where the IRGC trained them in the use of explosives; others were instructed as snipers. As a further incentive, MOIS offered $800 to Iraqis who killed American soldiers. The Iranian devices were said to account for half of the U.S. casualties, which reached 4431 by 2010.[10]

The sectarian mayhem was much worse. The Shiite militias took revenge on the Sunnis; they kidnapped, and killed their family members, burned their homes, and, in an early version of ethnic cleansing, stopped many from returning to their towns. The non-sectarian crime was also up. Murders, rapes, and property destruction and theft. Felony case dispositions, including torture, were rising rapidly, increasing by 40 percent between November 2003 and November 2004.[11]

In yet another technique honed in Lebanon known as "ballot and bullet," the Iranians pushed the Shiites to dominate the political process.

[10] Summer Basri, "Iran- Iraq: Can We Handle a Menace?" *NewsBlaze*, April 25, 2009, 9; Michael War, "Iran's Man in Iraq," *Time* April 12, 2006; Felter and Fishman, *Iranian Strategy in Iraq*, 34, 38; Kenneth R. Timmerman, Shadow Warriors: The Untold Story of Traitors, Saboteurs, and the Party of Surrender (Crown Forum, 2008), 72; Patrick Cockburn, Muqtada: Muqtada al-Sadr, the Shia Revival, and the Struggle for Iraq (Scribner, 2008), pp. 168–69; Michael Ware, "Inside Iran's Secret War for Iraq," *Time*, August 22, 2005; Iraq Coalition Military Fatalities By Year, www.icasualties.org.

[11] O'Hanlon et al., The State of Iraq: An Update, *The New York Times*, February 2005.

SCIRI, renamed the Supreme Islamic Council of Iraq (SICI), under the leadership of Abdul Aziz al-Hakim, who replaced his brother assassinated in 2003, played a leading role in this effort. Iran also counted on the Badr movement and the veteran Dawa party. To cover the northern part of the country, the IRGC-QF cultivated good relations with the Kurdish parties. The Iranians lavished large amounts of money on their preferred candidates while working behind the scenes to prevent the splintering of the Shiite vote. Special liaison personnel worked out of the Iranian embassy in Baghdad and consulates in Basra, Karbala, Irbil, and Sulaymaniyah. In fact, the Iranian diplomatic corps was a virtual extension of the QF; even the ambassador in Baghdad, Hassan Kazemi Qomi, was a former QF commander. In the words of one report, Iran worked to "create a weak federal state dominated by Shiites and amenable to Iranian influence."[12]

The QF used the political legitimacy which the Shiite bloc, the United Iraq Alliance, won in the 2005 election. Nouri al-Maliki, a virtually unknown politician from the Dawa Party, was picked to replace the disputed incumbent Prime Minister Ibrahim al-Jafari in 2006. The CIA, which interviewed several candidates, concluded that al-Maliki was the "most independent from Iran." In fact, Qassem Suleimani brokered the deal to nominate al-Maliki, a fact which was revealed much later. The QF chief also assured that former Shiite insurgents would find their way into positions in the security apparatus of the Interior and Defense Ministries.[13]

The Interior Ministry, under the control of SICI's Abdul Aziz al-Hakim, was notorious for inciting sectarian vengeance against the Sunnis. In 2006, John Pace, the United Nations human rights representative in Iraq, wrote that that "every month hundreds of Iraqis were being tortured to death or executed." Pace accused the ministry of "acting as a rogue element within the government." Other critics noted that "many of the 110,000 policemen and police commandos under the ministry's control are suspected of being former members of the Badr Brigade and the Mahdi Army who formed anti-Sunni death squads.[14]

Still recovering from its debacle about Saddam Hussein's alleged WMD, the CIA was caught by surprise by the scope of the Iranian involve-

[12] Michael Eisenstadt, Michael Knight, Ahmed Ali, "Iran's Influence in Iraq." Washington, DC: Washington Institute for Near East Policy, April 2011.

[13] Dexter Filkin, "The Shadow Commander," *New Yorker*, September 30, 2013.

[14] The Independent, February 6, 2006; Michael Eisenstadt, Michael Knight, Ahmed Ali, "Iran's Influence in Iraq." Washington, DC: Washington Institute for Near East Policy, April 2011.

ment in Iraq. Michael Rubin, a neoconservative scholar who had initially supported the invasion of Iraq, was one of first in the administration to realize the deep penetration of the IRGC-QF. Seconded from his job as the head of the Iran and Iraq desk in the Pentagon, to the Coalition Provisional Authority (CPA), Rubin quickly realized that the oil-rich south was becoming an "Iranian protectorate." More consequentially, as the QF settled in, the rest of the country was coming under the Iranian influence. In the words of one analyst, "the Quds Force … acted like an old-world mafia … keeping a close watch on local politicians who threatened Tehran's interests and objectives." Deeply perturbed, Rubin returned to the United States to take up a position at the American Enterprise Institute where he wrote about the growth of Iranian influence in the region.[15]

CIA reports implicated the IRGC-QF in the sectarian violence as well as the militias' attack on American soldiers. President Bush, already on the defensive about the missing WMD and facing reelection in 2004, was reluctant to admit to yet another miscalculation. The State Department issued a statement to the effect that there was "no hard evidence of any Iranian role in the unrest in Iraq." The normally outspoken Secretary of Defense Donald Rumsfeld tried to tone down the Iranian culpability. As he put it, "unless you physically see an [Electrostatic Discharge (ESD)] coming in an Iranian vehicle…it is very hard to tie a threat precisely to the government of Iran."[16]

Military analysts were less reticent to speak out. In February 2004, the Strategic Studies Institute at the Army's War College published a monogram titled, *The Strategic Implications of Intercommunal Warfare in Iraq*. Its author, W. Andrew Terrill, noted that "group identity which is critical throughout much of the Middle East, will provide a compelling context for regional bystanders" to ignite an ethnic and sectarian civil war in Iraq. Terrill pointed to Iran as the main player, writing that a "sectarian war would allow Iran" to seek to formulate policies that maximize its interest…while minimizing the influence of the U.S. and potential regional rivals such as Saudi Arabia. The author listed the challenges of avoiding a

[15] Jay Solomon, The Iran Wars. Spy Games, Bank Battles, loc. 1089, 1182.

[16] Rumsfeld, quoted in John Wobensmith, "Getting Smart on Iran," in Taking on Tehran: Strategies for Confronting the Islamic Republic, ed. Ilan Berman (Lanham, MD: Lexington Books, 2007), 7; quoted in Sasan Fayazmanesh, The United States and Iran: Sanctions, Wars and the Policy of Dual Containment (Routledge, 2008), 150.

catastrophic ethno-religious war, but was not very hopeful whether the United States would be successful.[17]

As it turned out, Terrill's pessimism was well justified. As noted above, by 2006, the sectarian strife had turned into a virtual civil war. Ayatollah Abdul Aziz al-Hakim played a highly duplicitous role. While encouraging the bloodbath, he reassured the world about Iraq's commitment to democracy. During a meeting with President Bush on December 4, 2006, al-Hakim stated: "we have gone a long way to establish a democratic and pluralistic society in Iraq." Al-Hakim reiterated that Iraq is "unified and strong" and described talk about sectarian strife as "an attempt to weaken the position in Iraq. Al-Hakim, who was touring the United States at the invitation of the American University's Center for Global Peace and the Catholic University of America, lectured on the "Freedom and Tolerance in Shiite Islam and the Future of Iraq."[18]

Al-Hakim's son, Amar Abdul Aziz al-Hakim, who established the al-Hakim Foundation, furthered the narrative of the Shiite liberalism and tolerance. Headquartered in Najaf, the Foundation, which had a consultative status in the critical United Nations Economic and Social Council, oversaw a vast network of educational, cultural, and religious institutions. But it was also a dispenser of the IRGC cultural products specializing in the Iraqi Shiite market. Amar al-Hakim's close relations with the Revolutionary Guards attracted the attention of the American intelligence. In February 2007, the American military arrested al-Hakim and his entourage when crossing the border from Iran on suspicion of smuggling weapons. He was speedily released when it became clear that SICI had organized several large demonstrations in Basra. Zalmay Khalilzad, the American ambassador to Baghdad, offered an apology to the al-Hakim family.[19] The al-Hakim incident was emblematic of the increasing American difficulty in dealing with the Iranians who pushed hard with their "Lebanonization" project. Unsurprisingly, criticism of the Bush administration policy became widespread. Peter W. Galbraith, an American diplomat attached to the United Nations, asserted that "George W. Bush's strategic gift to Iran

[17] W. Andrew Terrill, The Strategic Implications of Intercommunal Warfare in Iraq. Strategic Studies Institute, Army War College, Carlisle, PA, 2005, vi, 1, 19.

[18] President Bush Meets with His Eminence Abdul Aziz al-Hakim, Leader of the Supreme Council for the Islamic Revolution of Iraq, https://georgewbush-whitehouse.archives.gov/news/releases/2006/12/20061204-7.html#.

[19] Richard A. Oppel, Jr. "Son of Key Iraq Shiite Arrested at the Boarder," *New York Times*, February 23, 2007.

comprised the country's biggest gain in four centuries." He also pointed out that al-Maliki "was a hardline Shiite militant who had spent more than three decades in exile in Iran and Syria." Galbraith was one of the several analysts who urged to minimize Iran's gains by dividing Iraq into three sections; a Kurdish state, along with a Sunni, and a Shiite one. But the administration decided to surge the number of its troops in Iraq under the command of General David Petraeus who was widely considered an authority on counterinsurgency.[20]

As part of the new strategy, the U.S. funded the so-called Sunni Awakening, a coalition of Sunni tribes from Anbar province. Considered a response to the Iran-backed Shiites militias, the Sunni insurgents balanced the power of the al-Maliki government. The military surge and the empowering of the Sunnis had tamped down the sectarian tensions and restored a semblance of normalcy to Iraq. General Petraeus was so encouraged that he decided on a gradual withdrawal, to be completed by 2010.

Petraeus, who went on to head the Central Command (CENTCOM) in 2008, received wide acclaim for his role in Iraq. Not well understood at the time, however, was the fact that the IRGC-QF contributed to the relative peace. Known for its doctrine of strategic patience, the Guards decided to rein in their militias to fashion a more mainstream political position. In early March 2008, President Mahmoud Ahmadinejad arrived in Baghdad for a two-day visit, the first since the Iraq–Iran war. Ahmadinejad proclaimed the occasion to be a historic moment: "Today, by the grace of God, our two countries' leaders have agreed to cement their brotherly relations." Prime Minister al-Malaki ignored strong American pressure and welcomed the Iranian leader who also met Ayatollah Abdul Aziz al-Hakim.[21]

The IRGC-QF had additional ways to show the Americans their hold on Iraq. When in 2008, the al-Malaki government, prodded by the U.S, sent its military to Basra, the heart of the original Iranian "protectorate," the troops proved to be inept or refused to fight their Shiite compatriots. Al-Maliki was forced to negotiate a deal with Suleimani in Qom, a major humiliation for Washington. Moreover, as part of the deal, al-Maliki agreed to incorporate elements of the Badr Corps into the military, a decision which had negative implications for sectarian violence. To make his command of

[20] Peter W. Galbraith, Unintended Consequences. How the War in Iraq Strengthen American Enemies (New York: Simon & Schuster, 2008), 22, 115.

[21] Richard A. Opple Jr. and Ahmad Fadam, "Ahmadinejad, in Iraq, Chides Bush on Iran Criticism," *New York Times*, March 3, 2008.

Iraq clear, Suleimani sent a letter to General Petraeus reading: "Dear General Petraeus, you should know that I, Qassem Suleimani, control the policy for Iran concerning Iraq, Lebanon, Gaza, and Afghanistan. And indeed, the ambassador in Baghdad is a Quds Force member. The individual who's going to replace him is a Quds Force member." Indeed, Ryan Crocker, the American ambassador in Baghdad, was forced to meet with his Iranian counterpart, Hassan Kazemi Qomi in 2007 and 2008.[22]

Unlike the distant Lebanon, Iran could count on physical proximity to developing strong commercial ties with Iraq. The two countries signed numerous agreements to that effect. The total amount of Iran's non-oil exports to Iraq in 2003 was $436 million, mainly in food and household appliances. It surged to $4.6 billion in 2010, an astonishing 2200 percent increase.[23]

Interestingly, Iran provided financing for some of the trade. For instance, during his 2008 visit, Ahmadinejad agreed to extend a $1 billion loan to be paid back over 40 years for the purchase of Iranian goods and services.

The Guards invested in Iraq's infrastructure projects including power plants and improvement of the power grid. For instance, in 2007, the organization signed a $150 million contract to build a 300-MGW power plant in Baghdad, and it supplied power to Khanaqin, in Iraq's Diyala Governorate. It also sold 400-MGW electricity from its grid connected to lines that run between Marivan and Panjwin, and from the Iranian the port of Abadan to the Iraqi city of Alharasa. In June 2012, the MAPNA Group, an IRGC front company, commenced the construction of a 500-MGW Heidariya power plant in al-Haidiri district of Najaf, a 75 million-euro project which has supplied electricity to the cities of Najaf and Karbala.[24]

Drawing on its Al-Anbiya construction expertise, the IRGC launched major infrastructure projects. It signed a deal to build an airport near

[22] Dexter Filkins, The Shadow Commander, *The New Yorker*, September 30, 2013, https://www.newyorker.com/magazine/2013/09/30/the-shadow-commander; Afshon Ostovar, Vanguard of the Imam: Religion, Politics, and Iran's Revolutionary Guards (London: Oxford University Press, 2016); "Tehran Plays Double Game," *Newsweek*, July 23, 2007.

With the pliant al-Maliki in power, Suleimani could be excused for feeling like the "viceroy" of a Lebanonized Iraq.

[23] Iraq, *OEC*, atlas.media.mit.edu/en/profile/country/irq/#Imports.

[24] MNA, Ministry of Energy's Electric Gift to Iraq: Building Free Power Plant, *Mehr News Agency*, July 20, 2012, www.mehrnews.com/news/1653908/هدیه-برقی-وزارت-نیرو-به-ساخت-عراق-
نیروگاه-به-صورت-رایگان.

Najaf with a capacity to handle some 20,000 pilgrims a month. In 2009, an IRGC company won a $1.5 billion contract to build a new town, Haydar Ali, near Basra, and another company was commissioned to build a highway connecting Basra to several Iranian cities.

On April 12, 2009, the Iraqi authorities inaugurated an Iranian-built hospital in the Zafaraniyah district of Baghdad, operated by several Iranian doctors and medical teams. Hassan Kazemi Qomi, who was present at the ceremony, announced that his country would build more hospitals, power plants and schools to support the Iraqi people.[25]

Critics in Iran noted that the construction of this power plant came at a time when the Iraqi government had more than a $500 million outstanding debt for energy purchases. They also complained that the Iranian Ministry of Energy was in debt, owing more than $4 billion to private sector subcontractors. What incensed the critics most was the fact that the Guards would transfer electricity to Iraq while the Iranian grid struggled with shortages and blackouts.[26]

If the IRGC-QF, which sponsored many of the projects through its front companies, hoped to use the economic ties further its Lebanonization project, it was bound to be disappointed. For all his astuteness, Suleimani, who run Iraq as his private fiefdom, failed to realize that the sectarian strife created a powerful pushback from disaffected Sunni. Things came to a head in 2010 when al-Maliki narrowly lost the election to Ayad Allawi, a secular Shiite perceived as a Sunni supporter. After Allawi's attempt to lead a government failed due to the machination of Suleimani, al-Maliki had returned to the Prime Minister's office. While the QF celebrated, the Sunni backlash hit a tipping point, plunging the country into a new wave of bloodshed and chaos. Much to the surprise of the intelligence division of the IRGC, the new Sunni players emerged from the al-Qaeda Iraqi franchise, al-Qaeda in Iraq (AQI).

[25] AN, Iran will build a 220-Seat Hospital and 17 Schools in Iraq, *Aftab News*, April 12, 2009, aftabnews.ir/fa/news/86739/ایران-یک-بیمارستان-220-تختخوابی-و-17-مدرسه-در-عراق-میسازد; Tabnak, Iran will build 3000 Schools, April 29, 2012, *Tabnak*, www.tabnak.ir/fa/news/241223/سه-هزار-مدرسه-در-عراق-می-سازد-ایران.

[26] MNA, Ministry of Energy's Electric Gift to Iraq: Building Free Power Plant, *Mehr News Agency*, July 20, 2012, www.mehrnews.com/news/1653908/هدیه-برقی-وزارت-نیرو-به-عراق-ساخت-نیروگاه-به-صورت-رایگان.

IRGC and the Making of the Islamic State in Iraq

There was nothing in the initial actions of the Jordanian-born Musab al-Zarqawi, who was a low-ranking operative in al-Qaeda, to alarm the IRGC. To the contrary, the Revolutionary Guards allowed members of al-Qaeda who were fleeing the American advances in Afghanistan to transit or, on occasion, to shelter in Iran. Al-Zarqawi spent some time with Ansar al-Islam, an extreme Salafist group operating in Iraqi Kurdistan, but he soon launched his own franchise. The American invasion of Iraq was a significant boon to jihadist leaders looking for recruits to fight the Western infidels. Operating with a growing number of Iraqis who joined his group, al-Zarqawi managed to inflict some heavy casualties on the U.S.-led Coalition Forces and even the United Nations.

Initially, Suleimani welcomed Zarqawi who declared his allegiance to al-Qaeda, as an additional help in destabilizing Iraq and forcing the Americans out. In actuality, it was al-Zarqawi's hatred of the Shiites which fueled his murderous agenda. Scholars would later assert that, like other extreme Salafists, al-Zarqawi considered the Shiites to be takfiri, apostates whose blood could be shed. Although both bin Laden and Ayman al-Zawahiri disapproved of his attack on Shiites, the independently-minded jihadist leader rejected their concerns.

The intelligence division of the IRGC and the QF intelligence operatives in Iraq failed to realize that, unlike bin Laden, al-Zarqawi was influenced by the emerging Salafist eschatology to which he was exposed during a prison stint in Jordan. Among his jailhouse readings was *The Management of Savagery: The Most Critical State Through Which the Ummah Will Pass*, a book attributed to Mohammad Khalil al-Hakaymah. Writing under the name of Abu Bakr al-Naji, al-Hakaymah considered the American presence in Iraq to be part of the "end days." Other Salafi theologians asserted that, in addition to challenging the West, a final battle between the Sunnis and the Shiite takfiris would take place. To expedite this apocalyptic clash, al-Zarqawi resolved to use the savage methods that al-Naji advocated—assassination of leaders, suicide bombing of civilians, beheadings, attacks on holy shrines, and other atrocities—against the Shiites.

Eschatology aside, AQI took a dim view of the al-Maliki government mistreatment of the Sunnis. Al-Zarqawi considered the Shiite politician and his government to be Iranian stooges and Suleimani the real master of Iraq. Moreover, a popular Sunni conspiracy theory held that the United States invaded Iraq to transfer power to Iran. Emma Sky, a British expert who worked for General Raymond "Ray" Odierno, the commander of American forces in Iraq, noted that among Sunni Arabs "it was common to hear talk of secret deals between Iran and the United States, and lament that the U.S. gave Iraq to Iran."[27]

Even without the conspiracy theory, the Sunnis could point to real evidence of government abuse, mostly related to the infamous Special Investigative Directorate (SID) in the Ministry of Interior. SID, which was run by former members of the Shiite militias, was accused of mass torture and murder. The secret prison in Yadriya Street in Baghdad was a case in point. After an increasing number of Sunnis disappeared in 2004, their relatives complained to the American authorities. In 2005, U.S. soldiers raided the secret prison, one of the dozens of SID detention centers around the country, where they found emaciated and badly tortured prisoners. The Americans investigated the secret prisoners but did not publish the report so as not to inflame passions even further.[28]

Egregious as these methods were, reminding many of the brutal dictatorships of Saddam Hussein, the Sunnis were equally concerned about the quiet but effective campaign of ethnic cleansing. One of the newer tactics that the QF developed in Iraq and practiced later in Syria, it was conducted under the radar of the international community. Baghdad was particularly affected. The Sunni population declined from 45 percent in 2003 to 25 percent in 2007, making the capital a largely Shiite city.[29]

The Americans increased the pressure on al-Maliki, but could do little to limit the sectarian tensions. Emma Sky, a British expert on the Middle East who served in Iraq as the political advisor to U.S. General Ray Odierno from 2007 to 2010, explained that, in addition to deliberate policy, the Iraqi culture of tit-for-tat fueled violence. Reacting to the Yadriya Street "chamber of torture" and other atrocities, Al-Zarqawi embarked on a new round of violence: "through suicide bombing, kidnappings, assassinations and beheading [he] sought to provoke Shiite to retaliate against Sunnis." On orders of Suleimani, the head of the Badr Organization (formerly Badr Corps), Hadi al-Amiri, retailed by targeting Sunni officials and

[27] Emma Sky, quoted in Michael Weiss and Hassan Hassan, ISIS: Inside the Army of Terror (New York: Regan Arts), p. 64.

[28] Ned Parker, "Torture by Iraq Militias. The Report Washington Did Not Want You to See," *Reuters*, December 14, 2005.

[29] Sabrina Tavernise, District by District, Shiites Make Baghdad Their Own, *The New York Times*, December 23, 2006, https://www.nytimes.com/2006/12/23/world/middleeast/23shiites.html

rank and file. In response to the new wave of bloodshed, on February 22, 2006, a huge explosion severely damaged the Al-Askari Mosque in Samarra, one of the holiest Shiite shrines. Although AQI did not claim responsibility, violence against the Sunnis skyrocketed. The SID and the Shiite militias went into action leaving thousands of Sunnis, killed, tortured, or kidnapped, and dozens of Sunni mosques destroyed.[30]

In Baghdad alone, some 21 Sunni mosques were destroyed, and three Sunni Muslim clerics were shot dead. Some reports indicated that 379 Sunnis and 458 civilians were killed in Baghdad in February.[31]

The Petraeus surge and the Sunni Awakening stopped the violence, but some of the success should have been attributed to IRGC's decision to stand down and wait out the Americans. Pressured by Suleimani, al-Maliki negotiated a rather speedy departure date for the U.S. military. Under the terms of the U.S.–Iraq Status of Forces Agreement (SOFA), the Americans were to leave by December 31, 2011. The adjunct Strategic Framework Agreement (SFA) stipulated cooperation in several fields, including protection of human rights and minority status, a euphemism for ensuring the safety of the Sunnis and the Kurds. But the Iranian- sponsored militias had no intention to honor the SFA provisions and their attacks in Anbar province, the home of the Sunni Awakening, had rocketed.

In a well-trodden cycle of violence, the renewed assault on the Sunnis helped AQI under the leadership of Abu Omar al-Baghdadi who stepped in after Zarqawi was killed in 2006, to mobilize recruits. After Abu Omar was killed in 2010, his successor, Ibrahim Awad Ibrahim Ali Badr al-Samari, known by his nom de guerre as Abu Bakr al-Baghdadi, reshaped the organization and renamed it Islamic State of Iraq (ISI). According to terror experts, al-Baghdadi was sent to Bucca, an internment camp run by the American military in 2004 or 2005. In Bucca, he acquired a large following among fellow Salafists but also networked with former Baathist officers, some of whom were appointed to leading positions in ISI. The ISI chief also embraced a very strict reading of the *Management of Savagery* tactics as he prepared to battle the Shiites. "The Clanging of the Swords" series of propaganda videos, which the media-savvy al-Baghdadi produced, were extraordinarily vicious; they featured suicide bombing, beheading,

[30] Emma Sky, quoted in Michael Weiss and Hassan Hassan, ISIS: Inside the Army of Terror (New York: Regan Arts), p. 64.

[31] Wikileaks Shows Rumsfeld and Casey Lied about the Iraq War, *The Daily Beast*, February 21, 2011, http://www.thedailybeast.com/wikileaks-shows-rumsfeld-and-casey-lied-about-the-iraq-war.

and burnings of victims. Interestingly, however, as one expert noted, "most of the malice" was directed toward the Shiites.[32]

Al-Maliki helped to fuel the sectarian flames by purging the Sunni officials from the government. First, the bodyguards of Sunni ministers began to disappear. For instance, the moderate Finance Minister Rafi al-Issawi found that his bodyguards had been kidnapped. Feeling threatened, the minister himself escaped to Anbar province.[33]

With the last U.S. soldiers about to leave, on December 19, 2011, al-Maliki approved a warrant for the arrest of Tariq Hashimi, the vice president of Iraq and one of the highest-ranking Sunni politicians. Hashimi was accused of murdering Shiites, a charge he vehemently denied.[34]

The removal and arrest of the Sunni officials provoked large protests in Ramadi, Hawija and other Sunni towns where the populations harbored severe grievances against the Shiite-dominated government. Two issues created most of the resentment. The government failed to provide the Sunni cities with adequate services, including electricity, which was scarce in the post-invasion years. By ethnically prioritizing resources, al-Maliki was sending a strong signal that the Sunnis had become second-class citizens in their own country. Al-Maliki also refused to pay the salaries of many of the Sunni Awakening fighters. Worse, the government defaulted on hiring some 10,000 fighters, a promise that the Americans had made. In the end, only 17,000 were integrated into the Iraqi military, a fraction of the Shiite militiamen who wore the uniform. The position of the Awakening Council personnel was particularly precarious. Since they had fought al-Qaeda, al-Baghdadi people extracted their rather savage retribution. At the same time, al-Malaki prosecuted some Sunnis for anti-government activities from before they had joined the Council.[35]

Pushed to the edge, the Anbar Sunnis responded with a civil protest which occasionally turned violent. But things took a turn for the worse,

[32] Jessica Stern, "ISIS: The State of Terror," March 12, 2015, https://bi.hcpdts.com/reflowable/scrollableiframe/9780062395566.

[33] Martin Smith, The Rise of ISIS, www.pbs.org/wgbh/frontline/film/rise-of-isis/.

[34] Dan Morse, raq issues arrest warrant for vice president on terrorism charges, *The Washington Post*, December 19, 2011, https://www.washingtonpost.com/world/middle_east/iraqi-governmentissues-arrest-warrant-for-vice-president-on-terrorism-charges/2011/12/19/gIQA7bbD5O_story.html?utm_term=.83658bb42623

[35] Iraq's al-Maliki Seeks Arrest of Sunni VP as Terrorist, *Parliament in Uproar*, https://www.juancole.com/2011/12/iraqs-al-maliki-seeks-arrest-of-sunni-vp-as-terrorist-parliament-in-uproar.html.

when, on orders of Suleimani, several Shiite militias moved to quash the unrest. Sunni Sheikhs said the planned protest was to bring “honor of Iraqi female prisoners” back. Other Sunni provinces, Salahaddin and Mosul called for civil disobedience and expressed their support to Anbar’s protest against Maliki government, causing a considerable number of casualties.[36]

ISI seemed to be preparing for a future confrontation. In July 2012, al-Baghdadi executed the “Breaking of the Walls” project to free thousands of jailed Salafists. With the help of these men and an ever-expanding army of volunteers from all over the world, ISI launched a full-fledged operation in Anbar Province. His Bucca contacts paid off as, according to some accounts, members of some eighty Sunni tribes joined the fighting. These large numbers were critical because in January 2012 al-Baghdadi announced that a faction of ISI would cross into Syria to fight against Bashar al-Assad. Headed by Ahmad Hussein al-Shara, (a.k.a. Abu Mohammed al-Julani) his jail mate in Bucca, the new group took the name Jabhat al-Nusra. Reflecting its broader ambition, ISI morphed into Islamic State of Iraq and Syria (ISIS), also known as the Islamic State of Iraq and Levant (ISIL) or Islamic State (IS).

Buoyed by conquering a swath of territory in northern Syria, ISIS hoped to make the same strides victory in Iraq. With the Sunnis engaged in a growing protest movement across much of Iraq, the moment seemed opportune. In April 2013, a large demonstration took to the streets of al-Hawija, a town in Kirkuk province. Government forces responded with force, further inflaming passions among the Sunnis. Watching from the sidelines, al-Baghdadi decided on a full-fledged campaign. In early June 2014, ISIS occupied Mosul, the second largest city in Iraq. On June 29, 2014, al-Baghdadi announced the creation of the Islamic State, a new caliphate, naming himself as Caliph Ibrahim. The speed of al-Baghdadi’s progress was nothing short of astounding, especially as the much larger and better equipped Iraqi army virtually disintegrated. As one commentator put it, “the U.S. spent $25 billion in training and equipping the Iraq army for over eight years,” only to see it disintegrate “in a blink of an eye.”[37]

If the triumph of ISIS represented a blow to the United States, it was a virtual catastrophe for the IRGC-QF’s project of “Lebanonization” of Iraq and its chief architect, Qassem Suleimani.

[36] Reuters, Thousands protest against Iraqi PM’s policies in Anbar province, *Ammon*, December 28, 2012, en.ammonnews.net/article.aspx?articleNO=19763#.WfXUqdeWbIV.

[37] Martin Smith, The Rise of ISIS, www.pbs.org/wgbh/frontline/film/rise-of-isis/.

Toward a Shiite–Sunni–Kurdish Federation?

The fall of Mosul and the brutal rampage of al-Baghdadi's forces added tension to the fraught relations between the moderate normalizers and the hardliners in Tehran. Having struggled with a complicated nuclear negotiation, President Rouhani felt that the high-profile exploits of Suleimani called unnecessary attention to the continuation of Iran's revolutionary export. To minimize the danger to the pending nuclear agreement, the Joint Comprehensive Plan of Action (JCPOA), Rouhani wanted Suleimani replaced with Ali Shamkhani, advisor to the Supreme Leader and the secretary of the SNSC.

If the moderates acted logically in seizing upon the ISIS debacle to settle accounts with the QF, the criticism of Suleimani within the IRGC came as a surprise. As one observer noted, "ISIS was a stain on Suleimani who ignored them because he concentrated on defending Assad." Mashreq News, an outlet affiliated to the Revolutionary Guards, agreed with the moderates, writing that that Ali Shamkhani, secretary of the Supreme National Security Council of Iran (SNSC), and SNSC should make decisions regarding Iraq. In a remarkable piece of editorializing, Mashreq stated that "many do not pay much attention to the fact that the main mission of the Quds Force is fighting against Israel. Suleimani was put in charge of Iraq's political portfolio only because of the need to coordinate between the Iraq and the Syrian front operations." According to Mashreq, in the new division of labor, "Shamkhani will be in charge of Iraq and Suleimani will fulfill the decisions which will be made by SNSC."[38]

Suleimani, however, was not ready to give up his role as a "kingmaker" in Iraq politics. To the contrary, he insisted on supporting his long-term ally al-Maliki who was losing ground over his stringent anti-Sunni policies. With Sunni protests gaining strength, Rouhani concluded that al-Maliki needed to go. Ayatollah Khamenei apparently shared this view as well as Grand Ayatollah Sistani, the spiritual leader of the Iraq Shiites. The latter sent a message to al-Malaki urging him to resign, but Suleimani, who still pushed for a Lebanon version of Iraq, demurred. In Tehran, the moderate media amplified its attacks on Suleimani, blaming his brutal anti-Sunni campaign for the growth of ISIS. Backed by the Supreme Leader, Rouhani

[38] Mashreqnews, What is Behind Suleimani Removal from Being in Charge of Iraq Portfolio, *Mashreq News*, September 17, 2014, https://www.mashreghnews.ir/news/345951/پشت-پرده-شایعه-کنار-رفتن-سردار-سلیمانی-از-پرونده-عراق; Afshon Ostovar, Vanguard of the Imam: Religion, Politics, and Iran's Revolutionary Guards (London: Oxford University Press, 2016).

deputized Shamkhani to step in and engaged in the complex maneuvers to replace al-Maliki with the Heider al-Abadi, a moderate Shiite politician."[39]

Al-Abadi curtailed much of the abuse of the Sunnis and made his government more inclusive, a policy which President Rouhani supported. But Suleimani did not abandon the political-military arena. By utilizing his extensive contacts, the QF chief created the formidable Popular Mobilization Forces (PMF) or Hashid Shaabi, a conglomerate of several groups—Muqtada al-Sadr's Mahdi Army, Asa'ib Ahl al-Haq, Liwa'a Zulfiqar, Kata'ib Sayyid al-Shuhada, Kata'ib Hezbollah, Liwa'a 'Ammar Ibn Yasir and Munadhamat Badr. Saraya Khorasani, formed by the call of Ayatollah Khamenei, achieved a special prominence among the PMF. Its Iranian leader, Brigadier General Hamid Taghavi Abu Mariam who died in the fighting, was declared a war hero and featured on billboards together with Khamenei. The veteran operative Abu Mahdi al-Muhandis, second in command to Suleimani, served as the commander of the PMF and, a de facto deputy to Suleimani.[40]

The QF also collaborated with the Kurdish Regional Government (KRG) under Masoud Barzani which faced an assault of ISIS on its capital of Irbil in September 2015 Brigadier General Amir Ali Hajizadeh, the commander of the Guards Aerospace Force, told Iran TV that Suleimani played a key role in preventing the fall of Erbil. He noted that "if it were not because of the Iran's help, ISIS would have taken Kurdistan (Iraq). Sardar (General) Suleimani stood against ISIS with only 70 people and did not allow them to enter Erbil."[41]

With the Obama administration offering only tepid condemnations of ISIS, QF tried to bolster its credentials as the only defender of Iraq. Indeed, Suleimani was fond of emphasizing American weakness. Speaking on May 24, 2015, the QF chief was quoted as saying "Mr. Obama, you have not done a damned thing in Iraq." Al-Maliki, by then demoted to the

[39] DW, Quds Force Footprint in Iraq has become more Prominent, *Deutsche Welle*, www.dw.com/fa-ir/ردپای-سپاه-قدس-در-عراق-پررنگتر-شده-است/a-17830425.

[40] Ned Parker, Barak Dehgana Pish and Isabel Coles "How Iran's Military Chiefs Operate in Iran," *Reuters*, February 24, 2015, Abbas Qaidaari, Iran's new group in Iraq: Saraya al-Khorasani, *Al-Monitor*, January 11, 2015, www.al-monitor.com/pulse/originals/2015/01/iran-iraq-saraya-al-khorasani.html#ixzz4qQ3yVCaT.

[41] TNA, General Suleimani Blocked ISIS to Enter Erbil, *Tasnim News Agency*, September 24, 2014, https://www.tasnimnews.com/fa/news/1393/07/02/508313/ اگر-ایران-نبود-داعش-کردستان-عراق-را-می-گرفت-سردار-سلیمانی-با-70-نفر-جلوی-ورود-داعش-به-اربیل-را-گرفت; General Suleimani Saved Iraqi Kurdistan, *Aparat*, September 24, 2014, www.aparat.com/v/oluL6/ سردار_قاسم_سلیمانی_کر دستان_عراق_را_نجات_داد.

post of vice president, pushed the same theme. In his view, Iran was the sole supporter of Iraq in its hour of need.[42]

While concerned about the heavy losses in Iraq, the IRGC was encouraged by its popularity among Shiites and even some Sunnis who despaired of the unprecedented brutality of ISIS. Suleimani, assisted by the PMF, recaptured Amerli, Diyala governorate, on September 2, 2014, where he spoke about the "freedom of Khorramshahr," a reference to a famous battle during the Iran–Iraq war. The QF released pictures of Suleimani mingling with the liberators of Amerli to the Farsi media, which emphasized that the images were circulating on the Iraqi social media. For those who did not understand the message conveyed by the optics, the media emphasized that Suleimani was a true hero who saved many cities from the brutal rule of ISIS.[43]

Whatever the true level of their popularity, the Iranians did not rest on their military laurels alone. As before, the IRGC-QF used its soft power to win hearts and minds. On May 22, 2011, Iran signed an agreement with the Iraqi Ministry of Education to print more than 6 million textbooks, worth $2.2 million.[44] A year later, on April 29, 2012, Mohammad Reza Rahimi, a vice president in Ahmadinejad's government announced a contract to build 2000 residential units and 3000 schools building in Iraq.[45] Crucially, both the Iranian government and the IRGC's companies kept expanding the trade relations with Iraq to the tune of $12 billion in 2014, half of it non-oil related.[46]

[42] Bill Powell, Is Isis Winning? *Newsweek*, June 12, 2015; IP, "Bright Future Ahead of Iran-Iraq Relations, *The Iran Project*, January 5, 2017; Ramin Mostaghim and Patrick J. McDonnell, Iran slams U.S. over comments on Iraqi troops fighting Islamic State, *Los Angeles Times*, May 25, 2015, www.latimes.com/world/middleeast/la-fg-us-iran-islamic-state-20150525-story.html.

[43] YJC, What is the Secrete Behind Popularity of Quds Force Commander? *Young Journalist Club*, www.yjc.ir/fa/news/5595596/راز-محبوبیت-فرمانده-سپاه-قدس-در-عراق-چیست-تصاویر.

[44] Jamejam, Disclosure of Publishing Iraqi Textbooks in Iran, *Jamejam Online*, May 22, 2011, jamejamonline.ir/online/670950870671385424/افشاي-پرونده-چاپ-كتب-درسي-عراق-در-ايران.

[45] AN, Iran will build a 220-Seat Hospital and 17 Schools in Iraq, *Aftab News*, April 12, 2009, aftabnews.ir/fa/news/86739/مدرسه-در-عراق-میسازد-17-تختخوابی-و-220-ایران-یک-بیمارستان; Tabnak, Iran will build 3000 Schools, April 29, 2012, *Tabnak*, www.tabnak.ir/fa/news/241223/سه-هزار-مدرسه-در-عراق-می-سازد-ایران.

[46] Nasir al-Hassoun Iran-Iraq trade reached $12 billion in 2013 July 1, 2014, www.al-monitor.com/pulse/business/2014/07/iraq-iran-trade-increase-crisis-border-syria-jordan-turkey.html.

Several of the new trade ventures involved long-term projects. For instance, Iran Khodro, the largest car maker in Iran built six factories in Iraq, which were expected to produce some 20,000 cars a year. Also, the Iranians built a string of service facilities in large cities to service some 100,000 cars exported annually to Iraq.[47]

The IRGC-owned Arvand Free Trade Zone (FTZ) in the southern province of Khuzestan, turned it into a major export hub for goods destined for Iraq. By all accounts, the FTZ has been a success story, with some $176 million worth of commodities transiting in the six months of 2017 compared to $102 million in 2012; a 73 percent increase.[48]

Mixing security and business, a model that the IRGC adopted in Iran, proved successful, paving the way for security-only projects. Arguably, the land bridge to simplify the complex logistic of moving arms and supplies to Hezbollah, its chief link in the chain of revolutionary export proxies, was on the top of the agenda. There were three legs to the so-called Iran-Mediterranean land bridge. The Iranian part was planned to run from Tehran to Khosravi, a border crossing with Iraq. The Iraqi stretch was originally designed to traverse through Kurdish territory, but after Masoud Barzani, the head of the Kurdish Regional Government (KRG), objected the trail was moved to Sunni territory: Jalawla in Diyala province (south of Mosul) to Shirqat, Tal Afar, a mostly Sunni Turkman town north of Iraq before crossing into Syria through Mayadin. The Syrian part was planned to pass through Deir Al-Zor, Sukhna, Palmyra, and Damascus, before reaching the Lebanese border. To ensure control of the Iraqi stretch, the IRGC relied on the Harakat Hezbollah al-Nujaba, an Iraqi Shiite group under the command of Akram Kaabi, whose 10,000 fighters pledged allegiance to Ayatollah Khamenei.[49]

[47] Nasir al-Hassoun Iran-Iraq trade reached $12 billion in 2013 July 1, 2014, www.al-monitor.com/pulse/business/2014/07/iraq-iran-trade-increase-crisis-border-syria-jordan-turkey.html.

[48] FNA, Official: Iraq's Demand for Iranian goods High, *Fars News Agency*, August 20, 2013, en.farsnews.com/newstext.aspx?nn=13920529000395; FNA, Iran Exports $2.9bln Worth of Goods to Iraq in Six Months, *Fars News Agency*, October 21, 2013, en.farsnews.com/newstext.aspx?nn=13920729001260; IRNA, Iran exports $2.9bn worth of goods to Iraq in six months, *Irna News Agency*, October 21, 2013, en.irna.ir/News.aspx?Nid=80867654.

[49] Dexter Filkins, Iran Extends Its Reach in Syria, *The New Yorker*, June 9, 2017, https://www.newyorker.com/news/news-desk/iran-extends-its-reach-in-syria; Babak Dehghanpishesh, "The Iraqi Militia Helping Iran to Carve a Road to Damascus," *Reuter*, September 22, 2017.

With the Obama administration keeping a low profile in Iraq, the Revolutionary Guards were confident that they could force out ISIS and complete its Lebanonization project. Things changed, however, after Donald Trump's Pentagon chief, James Mattis, took a more activist stand against ISIS in Iraq. Mattis ordered the expediting of the training of the virtually defunct Iraq army and other security personnel. As of June 2017, the Americans reported training approximately 106,000 Iraqi security forces. Among the numbers were 40,000 regular troops, 15,000 police officers, 6,000 border guards, 21,000 Kurdish peshmerga—the military forces of the federal region of Iraqi Kurdistan—14,000 elite Counter-Terrorism Service personnel and another 9,500 "tribal mobilization forces."[50]

Pentagon officials and their civilian colleagues in the American embassy in Baghdad expressed hope that the new professional security force would provide a counterbalance to Suleimani's PMF. They also hoped to avoid the type of Shiite Sunni tensions Iraq had experienced in the 2000s. But in 2016, prodded by Suleimani, the Iraqi parliament passed a bill recognizing the PMF as a government entity operating alongside the military.[51]

While the bill was a victory for the IRGC-QF, obstacles to Iranian domination had multiplied. As Chap. 7 discusses in greater detail, following a decision to stand up more forcefully to Iran, Saudi Arabia promised considerable investment to rebuild the mostly Sunni cities in Anbar province—Mosul, Ramadi, and Fallujah. The Saudi government considered the reconstruction of the country to be an essential step into returning Iraq to the "Arab folds," a euphemism for pushing back on Iran's influence.[52]

The Americans supported the Saudi plan and offered some innovative solutions to revitalize the Iraq economy, still heavily dependent on oil. They persuaded the al-Abadi government to turn Root 1, a major artery from Baghdad to Amman, Jordan, into a toll road. The road advocates pointed out that the new highway would revitalize trade in the depressed Sunni province while protecting it from the Shiite militias and bandits who preyed on the traffic. The government awarded the contract to the Olive

[50] DM, Mosul victory marks win for Pentagon training plan, *Daily Monitor*, July 10, 2017, www.monitor.co.ug/News/World/Mosul-Pentagon-training-plan-jihadists-US-military-/688340-4008060-ukwo4uz/index.html.

[51] Tim Arango, "Iran Dominates in Iraq after U.S. Handed the Country Over," *New York Times*, July 15, 2017.

[52] Martin Chulov, Saudis in Talk Over the Alliance to Rebuild Iraq and Return it to the "Arab Fold." *Guardian*, August 18, 2017.

Group, a private U.S.-based security group, which promised to carry out repairs, install road rests and provide security. The Sunnis were highly supportive of the project, but unsurprisingly, the IRGC-QF and some Shiite politicians were bitterly opposed. For instance, Izzat al-Shahbander, a prominent Shiite leader, warned that the militias are stronger than the Iraqi army and would not allow the Americans to operate the tolls. Al-Shahbander threatened that Iran would work to remove al-Abadi.[53]

Still, the IRGC-QF may have overestimated its influence, because the ISIS calamity left a deep scar on the body politic of Iraq. Even hardliners such as Muqtada al-Sadr had second thoughts about pushing the Sunnis into the arms of ISIS, a scenario that they would not want to see repeated. Interestingly, the former leader of the radical Mahdi Army seemed to accept the view that rehabilitation of the Sunnis is essential for Iraq. On a July 30, 2017 visit to Riyadh, al-Sadr expressed support for closer relations with the kingdom. Iran hardliners described Sadr's visit as "a step that hurts the feelings of Muslims" and predicted that the Saudi overture would lead to the weakening of Shiites in Iraq. Specifically, the Revolutionary Guards accused al-Sadr of undermining the influence of the PMF and, by implication, Iran. The Rouhani pragmatists, however, lauded the trip, stating that "Sadr's visit to Saudi Arabia is very important given the tense relations in the region."[54]

The unprecedented rapprochement between Riyadh and Baghdad could have spelled the end of IRGC-QF's Lebanonization project. But the Kurdish referendum of September 22, 2017, gave the scheme a new lease of life. As expected, a clear majority of the Kurds voted in favor of independence, a ballot which both the Rouhani government and the Guards condemned. Indeed, the Kurdish referendum inspired the Sunnis to call for autonomy, a development which could have led to a federal Iraq and a much-shrunken Iranian influence. To stave off such an event, Suleimani urged Prime Minister Abadi to respond forcefully. After issuing an ultimatum to the Kurdish Regional Government (KPG), on October 16, three weeks after the referendum, the Iraqi forces helped by the PMF, invaded the disputed area of Kirkuk. The head of the KRG Massoud Barazani vowed that the peshmerga forces would fight to protect the territory, but most withdrew without offering resistance.

[53] Tim Arango, "U.S. Sees a Vital Iraqi Toll Road, but Iran Sees a Threat," *New York Times*, May 27, 2017.

[54] Donyaye Eghtesad, Translation of Muqtada Al-Sadr to Saudi Arabia, *Donyaye Eghtesad Newspaper*, August 1, 2017, No. 4107, p. 2.

Subsequent accounts revealed that Suleimani, who did not hide his presence in the area, orchestrated the withdrawal of the peshmerga fighters loyal to the family of Jalal Talabani, the head of Patriotic Union of Kurdistan (PUK) and a rival of the Barzani clan. The Talabani faction, rumored to be led by Jalal's son, Bafel, had enjoyed good relations with the QF and were quite eager to torpedo Barazani's independence project. In yet another blow to the Kurdish aspirations, the PMF took the strategic crossing of Faiysh Khaour on the border with Syria, the only free land passage out of Kurdish Iraq. Baghdad had previously closed the airspace to airports in Sulaymaniyah and Erbil.

Widely blamed for his miscalculation, Barazani resigned, raising questions about the future of the KRG and indeed Iraq. The central government in Baghdad ruled out a formal federation, either with the Kurds or the Sunnis, a stand which the IRGC-QF has supported. While savoring their Kirkuk victory, Suleimani and the IRGC chief Major General Ali Jafari have been treading carefully, undoubtedly because of the backlash from the United States. As already noted, the Trump administration has prepared new sanctions against the organization, and members of Congress pressured Abadi to dissociate himself from the PMF. Equally important, the Rouhani team has cautioned the Guards against an aggressive drive towards Lebanonization. Iraq is only one of several hotspots where the high-profile presence of the IRGC-QF clashed with Rouhani's vision of normalization. If anything, the Guards involvement in Syria has been even more detrimental, which is discussed the next chapter.

Conclusion

For the Islamic Republic, Iraq represented a perfect terrain for spreading the Islamic revolution. Replacing the Baathist regime with the Islamic Republic of Iraq was nothing short of a miracle. From the perspective of the Iranians, there was no shortage of Iraqi Shiites who were willing to play a role in this venture.

Indirectly, 9/11 turned out to be the answer which the revolutionary exporters were waiting. Interestingly enough, for U.S. neoconservatives who had little understanding of the reality in Iraq, regime change in Baghdad was just part of a larger plan to undermine Iran's bourgeoning hold on the Middle East. It was only after the invasion in March 2003 that the Americans learned of the extent of the Iranian involvement in Iraq.

The removal of Saddam Hussein enabled the IRGC-QF to move swiftly to expand its influence, and pushed the Shiites to dominate the political process, a policy that led to a notorious sectarian vengeance against the Sunnis.

The American miscalculation, effectively a strategic gift to Iran, led to the country's biggest gain in four centuries, to the extent that some officials in the Bush administration suggested dividing Iraq into three sections to minimize Iran's gains. The idea was rejected by President Bush in favor of opting for a new strategy of Sunni Awakening, a strategy that had, to some extent, tamped down the sectarian tensions and restored a semblance of normalcy to Iraq.

The sectarian strife, mistreatment of the Sunnis, ethnic prioritizing of resources and the arrest of Sunni officials by al-Maliki government sent a strong signal that the Sunnis had become second-class citizens in their own country, creating a powerful pushback from disaffected Sunnis and paving the way to the formation of Sunni radical groups.

It was only the emergence of ISIS that represented a virtual catastrophe for the IRGC-QF's project of "Lebanonization," which added tension to the fraught relations between the moderate normalizers and the hardliners in Tehran. Having struggled with a complicated nuclear negotiation, President Rouhani felt that the high-profile exploits of Suleimani called unnecessary attention to the continuation of Iran's revolutionary export.

Obstacles to Iranian domination had multiplied following a Trump's coming to power and taking a more activist stand against ISIS in Iraq, and after Saudi Arabia's decision to stand up more forcefully to Iran, by promising considerable investment to rebuild the mostly Sunni cities as an essential step to return Iraq into the "Arab folds," a plan which was supported by the United States.

Still, the IRGC-QF may have overestimated its influence, because the ISIS calamity left a deep scar on the body politic of Iraq. Even hardliners such as Muqtada al-Sadr seemed to accept the view that rehabilitation of the Sunnis is essential for Iraq. The unprecedented rapprochement between Riyadh and Baghdad could have spelled the end of IRGC-QF's Lebanonization project. But the Kurdish referendum gave the scheme a new lease of life. Equally important, the Rouhani team has cautioned the Guards against an aggressive drive towards Lebanonization. Iraq is only one of several hotspots where the high-profile presence of the IRGC-QF clashed with Rouhani's vision of normalization.

References

Abbas Qaidaari, Iran's new group in Iraq: Saraya al-Khorasani, *Al-Monitor*, January 11, 2015, www.al-monitor.com/pulse/originals/2015/01/iran-iraq-saraya-al-khorasani.html#ixzz4qQ3yVCaT.

Afshon Ostovar, Vanguard of the Imam: Religion, Politics, and Iran's Revolutionary Guards (London: Oxford University Press, 2016).

Aftab, Iran will build a 220-Seat Hospital and 17 Schools in Iraq, *Aftab News*, April 12, 2009, aftabnews.ir/fa/news/86739/بیمارستان-یک-ایران-220-و-تختخوابی-17-مدرسه-در-عراق-میسازد.

Andrew Terrill, The Strategic Implications of Intercommunal Warfare in Iraq. Strategic Studies Institute, Army War College, Carlisle, PA, 2005.

Aparat, General Suleimani Saved Iraqi Kurdistan, *Aparat*, September 24, 2014, www.aparat.com/v/oluL6/سردار_قاسم_سلیمانی_کردستان_عراق_را_نجات_داد.

Aram Rostam, The Man Who Pushed America to War: The Extraordinary Life, Adventures, and Obsessions of Ahmad Chalabi.

Babak Dehghanpishesh, "The Iraqi Militia Helping Iran to Carve a Road to Damascus," *Reuter*, September 22, 2017.

Bill Powell, Is Isis Winning? *Newsweek*, June 12, 2015.

David Wurmser, Tyranny's Ally (Washington, DC: American Enterprise Institute, 1999); quoted in Aram Rostam.

Defa, Formation of Badr Army, the Factor for Successful Operations Against Baath Regime, defapress.ir/fa/news/26472/تشکیل-سپاه-بدر-عامل-یکپارچگی-عملیات-نظامی-علیه-رژیم-بعث.

Dexter Filkins, Iran Extends Its Reach in Syria, *The New Yorker*, June 9, 2017, https://www.newyorker.com/news/news-desk/iran-extends-its-reach-in-syria.

Dexter Filkins, The Shadow Commander, *The New Yorker*, September 30, 2013, https://www.newyorker.com/magazine/2013/09/30/the-shadow-commander.

DM, Mosul victory marks win for Pentagon training plan, *Daily Monitor*, July 10, 2017, www.monitor.co.ug/News/World/Mosul-Pentagon-training-plan-jihadists-US-military-/688340-4008060-ukwo4uz/index.html.

Donyaye Eghtesad, Translation of Muqtada Al-Sadr to Saudi Arabia, *Donyaye Eghtesad Newspaper*, August 1, 2017, No. 4107.

DW, Quds Force Footprint in Iraq has become more Prominent, *Deutsche Welle*, www.dw.com/fa-ir/ردپای-سپاه-قدس-در-عراق-پررنگتر-شده-است/a-17830425.

Emma Sky, quoted in Michael Weiss and Hassan Hassan, ISIS: Inside the Army of Terror (New York: Regan Arts).

FNA, Iran Exports $2.9 bln Worth of Goods to Iraq in Six Months, *Fars News Agency*, October 21, 2013, en.farsnews.com/newstext.aspx?nn=13920729001260.

FNA, Official: Iraq's Demand for Iranian goods High, *Fars News Agency*, August 20, 2013, en.farsnews.com/newstext.aspx?nn=13920529000395.

Global Security, *Badr Corps*, www.globalsecurity.org/military/world/para/badr.htm.

I Worship the God of Ayatollah Sistani, *Bashgah Andishe*, January 27, 2004, www.bashgah.net/fa/content/show/5637.

IP, “Bright Future Ahead of Iran-Iraq Relations,” *The Iran Project*, January 5, 2017.

Iraq Coalition Military Fatalities by Year, www.icasualties.org.

Iraq, *OEC*, atlas.media.mit.edu/en/profile/country/irq/#Imports.

Iraq’s al-Maliki Seeks Arrest of Sunni VP as Terrorist, *Parliament in Uproar*, https://www.juancole.com/2011/12/iraqs-al-maliki-seeks-arrest-of-sunni-vp-as-terrorist-parliament-in-uproar.html.

IRNA, Iran exports $2.9 bn worth of goods to Iraq in six months, *Irna News Agency*, October 21, 2013, en.irna.ir/News.aspx?Nid=80867654.

Jamejam, Disclosure of Publishing Iraqi Textbooks in Iran, *Jamejam Online*, May 22, 2011, jamejamonline.ir/online/670950870671385424/افشاي-پرونده-چاپ-كتب-درسي-عراق-در-ايران.

Jay Solomon, The Iran Wars. Spy Games, Bank Battles.

Jessica Stern, “ISIS: The State of Terror,” March 12, 2015, https://bi.hcpdts.com/reflowable/scrollableiframe/9780062395566.

Joseph Felter and Brian Fishman, Iranian Strategy in Iraq: Politics and “Other Means” (West Point, NY: Combating Terrorism Center, West Point, 2008).

Kenneth R. Timmerman, Shadow Warriors: The Untold Story of Traitors, Saboteurs, and the Party of Surrender (Crown Forum 2008).

Khamenei.ir, Travel to Syria, *Khamenei.ir*, September 6, 1984, farsi.khamenei.ir/print-content?id=8699.

Martin Chulov, “Saudis in Talk Over the Alliance to Rebuild Iraq and Return it to the Arab Fold,” *Guardian*, August 18, 2017.

Martin Smith, The Rise of ISIS, www.pbs.org/wgbh/frontline/film/rise-of-isis/.

Mashreqnews, What is Behind Suleimani Removal from Being in Charge of Iraq Portfolio, *Mashreq News*, September 17, 2014, https://www.mashreghnews.ir/news/345951/پشت-پرده-شایعه-کنار-رفتن-سردار-سلیمانی-از-پرونده-عراق.

Michael Eisenstadt, Michael Knight, Ahmed Ali, “Iran’s Influence in Iraq.” (Washington, DC: Washington Institute for Near East Policy, April 2011).

Michael War, “Iran’s Man in Iraq,” *Time*, April 12, 2006.

Michael Ware, “Inside Iran’s Secret War for Iraq,” *Time*, August 22, 2005.

MNA, Ministry of Energy’s Electric Gift to Iraq: Building Free Power Plant, *Mehr News Agency*, July 20, 2012, www.mehrnews.com/news/1653908/هدیه-برقی-وزارت-نیرو-به-عراق-ساخت-نیروگاه-به-صورت-رایگان.

Nasir al-Hassoun Iran-Iraq trade reached $12 billion in 2013 July 1, 2014, www.al-monitor.com/pulse/business/2014/07/iraq-iran-trade-increase-crisis-border-syria-jordan-turkey.html.

Ned Parker, "Torture by Iraq Militias. The Report Washington Did Not Want You to See," *Reuters*, December 14, 2005.

Ned Parker, Barak Dehgana Pish and Isabel Coles "How Iran's Military Chiefs Operate in Iran," *Reuters*, February 24, 2015.

Newsweek, "Tehran Plays Double Game," *Newsweek*, July 23, 2007.

O'Hanlon et al., The State of Iraq: An Update, *The New York Times*, February 2005.

Patrick Cockburn, Muqtada: Muqtada al-Sadr, the Shia Revival, and the Struggle for Iraq (Scribner, 2008).

Peter W. Galbraith, Unintended Consequences. How the War in Iraq Strengthen American Enemies (New York: Simon & Schuster, 2008).

President Bush Meets with His Eminence Abdul Aziz al-Hakim, Leader of the Supreme Council for the Islamic Revolution of Iraq, https://georgewbush-whitehouse.archives.gov/news/releases/2006/12/20061204-7.html#.

Ramin Mostaghim and Patrick J. McDonnell, Iran slams U.S. over comments on Iraqi troops fighting Islamic State, *Los Angeles Times*, May 25, 2015, www.latimes.com/world/middleeast/la-fg-us-iran-islamic-state-20150525-story.html.

Reuters, Thousands protest against Iraqi PM's policies in Anbar province, *Ammon*, December 28, 2012, en.ammonnews.net/article.aspx?articleNO=19763#.WfXUqdeWbIV.

Richard A. Oppel, Jr. "Son of Key Iraq Shiite Arrested at Iran's Border," *New York Times*, February 23, 2007.

Richard A. Opple Jr. and Ahmad Fadam, "Ahmadinejad, in Iraq, Chides Bush on Iran Criticism," *New York Times*, March 3, 2008.

Rumsfeld, quoted in John Wobensmith, "Getting Smart on Iran," in Taking on Tehran: Strategies for Confronting the Islamic Republic, ed. Ilan Berman (Lanham, MD: Lexington Books, 2007), 7.

Sasan Fayazmanesh, The United States and Iran: Sanctions, Wars and the Policy of Dual Containment (Routledge, 2008).

Speech in Friday Prayers, *Khamenei.ir*, April 11, 2003, farsi.khamenei.ir/speech-content?id=3168.

Summer Basri, "Iran–Iraq: Can We Handle a Menace?" *NewsBlaze*, April 25, 2009, 9.

Tabnak, Iran will build 3000 Schools, April 29, 2012, *Tabnak*, www.tabnak.ir/fa/news/241223/ایران-سه-هزار-مدرسه-در-عراق-می-سازد.

Tim Arango, "Iran Dominates in Iraq after U.S. Handed the Country Over," *New York Times*, July 15, 2017.

Tim Arango, "U.S. Sees a Vital Iraqi Toll Road, but Iran Sees a Threat," *New York Times*, May 27, 2017.

Tim Arango, "U.S. Sees a Vital Toll Road, but Iran Sees a Threat," *New York Times*, May 27, 2017.

TNA, General Suleimani Blocked ISIS to Enter Erbil, *Tasnim News Agency*, September 24, 2014, https://www.tasnimnews.com/fa/news/1393/07/02/508313/اگر-ایران-نبود-داعش-کردستان-عراق-را-می-گرفت-سردار-سلیمانی-با-70-نفر-جلوی-ورود-داعش-به-اربیل-را-گرفت.

Wikileaks Shows Rumsfeld and Casey Lied about the Iraq War, *The Daily Beast*, February 21, 2011, http://www.thedailybeast.com/wikileaks-shows-rumsfeld-and-casey-lied-about-the-iraq-war.

YJC, What is the Secrete Behind Popularity of Quds Force Commander? *Young Journalist Club*, www.yjc.ir/fa/news/5595596/ -عراق-در-قدس-سپاه-فرمانده-محبوبیت-راز چیست-تصاویر.

CHAPTER 6

Iran and Syria: Leveraging the Victory?

By planning to use the Shiites as agents of the revolution, the new regime did not take the daunting Shiite demographics into full account. The only sizable Shiite group lived in Lebanon, a non-contiguous state, leaving Mustafa Chamran and Mohammed Montazeri (the son of Ayatollah Hossein Ali Montazeri) from the Office of Liberation Movement (OLM) scrambling for options. They were forced to appeal to Hafez Assad, the Syrian dictator, for help. Although Assad described Syria as the "beating heart of Arabism," his impoverished country needed the oil-rich regime in Tehran. Also, as an Alawite, a minority sect embedded in the majority Sunni population, Assad was more inclined to deal with his unpopular Shiite neighbor.

Ayatollah Ali Meshkini, who had extensive contacts in Syria, and Ali Akbar Mohtashami-Pur, the Iranian ambassador to Damascus, negotiated a deal with Assad who allowed the Iranians to transfer arms to the fledgling military organization Hezbollah. Assad also permitted the Islamic Revolutionary Guards Corps (IRGC) to construct a training facility for Hezbollah fighters in Zabadani, near Damascus. Over the years, the relations between Syria and the regime had become closer. During the Iran–Iraq war, Assad defied the Arab states that supported Saddam Hussein by turning off the Iraqi–Syrian oil pipeline. When the United States imposed an arms embargo on Iran to help Saddam Hussein, Damascus became a conduit for illicit arms. Two decades later,

F. Rezaei, *Iran's Foreign Policy After the Nuclear Agreement*, Middle East Today, https://doi.org/10.1007/978-3-319-76789-5_6

Syria helped the regime to evade the international sanctions imposed to roll back Iran's nuclear project.[1]

Iran's strategic doctrine indicated that the relations between Tehran and Damascus went beyond a traditional alliance between two countries. The IRGC understood the depth of resentment that its revolutionary export had created among the Sunni Arab states. To wit, the IRGC commanders took to referring to this hostile environment as a "swamp" for which they needed some "stepping stones." In time, Syria was elevated from a "stepping stone" to a virtual territorial extension of Iran known as the "35th province." On February 14, 2013, Hujjat al-Islam Mahdi Taeb, the head of the Ammar Strategic Base, an organization established to promote "soft war tools" and an adviser to the Supreme Leader, declared that Syria's strategic importance exceeds that of Khuzestan province in southern Iran.[2]

Hassan Abbassi, the head of the Center for the Borderless Security Doctrinal Analysis, added that "compared to Iraq, Lebanon, and Yemen, Syria is a priority. If one of these countries fall, we will not lose more than what we will lose if Syria falls. To preserve the achievements of the Islamic Revolution, we should have stepped into Syria. No wise man waited to lose his capital before his eyes."[3]

Ali Akbar Velayati, Khamenei's foreign policy representative, asserted that "Syria is a golden ring of resistance to Zionism. Iran supports it, because if Syria falls and its government collapses…it will affect Lebanon and Iraq. The resistance line begins with Iran, and passes through Iraq to Syria, Lebanon, and Palestine."[4]

General Mohammad Ali Jafari, the IRGC chief, explained that Syria's vicinity to "occupied Palestine" and its proximity to Iraq added to its strategic importance. Jafari also found the Assad family to be in tune in

[1] Ali Alfone, Iran Unveiled: How the Revolutionary Guards Is Transforming Iran from Theocracy into Military Dictatorship (AEI Press, 2013), 221–226.

[2] Asriran, Chief of Amar Base, Syria is the 35th province of Iran, *Asre Iran*, February 14, 2013, http://www.asriran.com/fa/news/257730/رئیس-قرارگاه-عمار-سوریه-استان-سی-و-پنجم-است-اگر-دشمن-بخواهد-سوریه-یا-خوزستان-را-بگیرد-اولویت-حفظ-سوریه-است-تحریم-ها-مثل-قبل-نیست.

[3] Hassan Abbassi, Syria: Islamic World's Strategic Depth, in Mohammad Khalilpour, A Different Story of Logic of Being in Syria (Tehran: Masir-e Aftab, 2017), pp. 57–74.

[4] Ali Akbar Velayati, Syria is the Golden Ring of Resistance line, in Mohammad Khalilpour, A Different Story of Logic of Being in Syria (Tehran: Masir-e Aftab, 2017), pp. 103–109.

defending the revolution, promoting Islamic resistance, and taking an uncompromising stand against Israel.[5]

The special role that Syria played in the IRGC strategic doctrine was not lost on observers. As one of them put it, "gradually over three decades, Syria became a cornerstone of Islamic Republic." It was during this period that the power balance between Damascus and Tehran had changed. Whereas at the beginning of the alliance Syria controlled the relationship, by the early 2000s, Iran became the dominant partner. Bashar al-Assad who took over from his father on July 17, 2000, was an untested replacement to the heir apparent, his older brother Bassel al-Assad who died in a car accident on January 21, 1994. Inner IRGC deliberations hinted at the need to prop up the younger Assad amid fears that the Sunni majority would raise up against its Alawite masters.[6]

With Bashar safely enthroned, the contacts between Syria and Iran became routine, a policy that intensified in the wake of the American invasion of Iraq. In July 2004, Assad made his third state visit to Tehran, and three months later President Mohammad Khatami traveled to Syria. U.S. Middle East policy and the Iraq war were high on the agenda of their bilateral talks, and the two allies decided to present a "united front" against the regional challenges. Of course, bailing out Assad was not cheap, but the payoff was considerable. In the words of one observer, Iran gained "military bases, land routes, airports, and port facilities."[7]

Real estate aside, Iran's alliance with Syria gave the Shiite regime a measure of legitimacy in the Sunni Arab world. In fact, the IRGC was quick to claim that the Assad dynasty was a pillar of the axis of resistance against the United States, Israel, and the Arab regimes that collaborate with Washington. As Hassan Abbassi noted, "Wherever there is a conflict that the Zionist regime is on one side, Iran will be on the other side fighting Israel along with that oppressed nation. We must point out that we will not leave the Syrian people and its government alone against the Zionist regime and its allies."[8]

[5] Entekhab, General Jafari: Bashar Assad is not the Reason for War in Syria, its Resistance and Islamic Revolution, *Entekhab*, September 27, 2017, www.entekhab.ir/fa/news/367339/
سردار-جعفری-هدف-جنگ-در-سوریه-بشار-اسد-نیست-مقاومت-و-انقلاب-اسلامی-است.

[6] Afshon Ostovar, Vanguard of the Imam: religion, politics, and Iran's Revolutionary Guards (Oxford University Press, 2016).

[7] Ibid.

[8] Hassan Abbassi, Syria: Islamic World's Strategic Depth, in Mohammad Khalilpour, A Different Story of Logic of Being in Syria (Tehran: Masir-e Aftab, 2017), pp. 57–74.

After President George W. Bush included Iran in the "axis of evil" in his address in January 2002, the hardline Iranian press responded by touting the virtues of the "axis of resistance." Indeed, the IRGC-QF described its involvement in Iraq and Syria as the broadening of the "axis of resistance." A few years later, Ali Akbar Velayati, speaking for the Supreme Leader stated that "the chain of resistance against Israel by Iran, Syria, Hezbollah, the new Iraqi government and Hamas passes through the Syrian highway…Syria is the golden ring of the chain of resistance against Israel."[9]

So long as the Assad dynasty was firmly in control of the country, the "golden ring" carried relatively a low maintenance costs for Iran. Both the moderate normalizers and the hardline Principalists were equally supportive of the extensive strategic links with Damascus. But things became more complicated when Ahmadinejad, denounced for his anti-Semitic and anti-Israeli rhetoric, moved Iran closer to a pariah state. Assad found himself in a similar position. Accused of being complicit in the murder of the Lebanese Prime Minister Rafiq Harari in 2005, Syria was pressured to cooperate with a UN investigation. The Bush administration imposed sanctions on Syria, which led Assad to withdraw his troops from Lebanon. Iran expressed its support for Syria, and Ahmadinejad visited Damascus on January 19, 2006. He was received with great honor by Assad who lavished praise on his "best friend," and the two proclaimed their desire to strengthen the "axis of resistance." At a joint press conference, both leaders emphasized the theme of "resistance" against the United States, Israel and the West. A Western commentator noted with some astonishment that the "Iranians [had] exited the mainstream" and set "in motion a regional and global game plan with the audacious goal of laying claim to the ownership of revolutionary Islamism."[10]

As noted in Chap. 1, the IRGC-QF and the Supreme Leader were excited about the Arab Spring. Ayatollah Khamenei predicted that, like with the fall of Saddam Hussein, the revolt against the "infidel" rulers—his term for the Arab heads of state—would further the goal of the regime.

[9] Ali Akbar Velayati, Syria is the Golden Ring of Resistance line, in Mohammad Khalilpour, A Different Story of Logic of Being in Syria (Tehran: Masir-e Aftab, 2017), pp. 103–109.

[10] David M. Smick, Iran's Ahmadinejad: Crazy or Crazy Like a Fox? *The International Economy* Winter 2006; Payvand, Defense Minister: Iran, Syria defense cooperation at highest level, June 16, 2006, www.payvand.com/news/08/may/1278.html; AFP, Iran, Syria sign defense agreement, June 16, 2006, https://www.ynetnews.com/articles/0,7340,L-3263739,00.html; The Result of Iran's President's Visit to Syria, January 20, 2006, www.nourizadeh.com/archives/001640.php#more.

The hardliners expected the "constructive chaos" of the popular revolts to further the ethnic and religious divisions on which the revolution had thrived. Understandably, however, after making a huge investment in the Assad dynasty, the IRGC-QF was not willing to tolerate this "constructive chaos" in Syria.[11]

From Arab Spring to Civil War

After watching the successful popular protest in other Arab countries, the Syrians took to the streets as well. On the advice of Qassem Suleimani, the government responded with considerable violence against demonstrators in Damascus, Aleppo, and other cities. According to the Iranian media, the IRGC supplied "coercive means" to Assad, a reference to a contingent of Basij troops sent to help the Syrian police and paramilitary units disperse the protestors. The IRGC leaders were confident that the Basij, which crushed the Iranian Green Movement in 2009, would be able to put down the rebellion in Syria. Testifying to the importance of the mission, the IRGC-QF sent several high-ranking commanders. Among them were Brigadier Mohsen Chizari, head of the organization's operation and training, Ismail Ahmadi-Moghaddam, Iran's Chief of Police, Brigadier General Hossein Ahstari, the commander of Law Enforcement Force of Islamic Republic of Iran (LEF) and his deputy Ahmad Reza Radan. For his part, Assad deployed the feared Shabiha ("ghosts"), violent Alawite gangs that brutalized protestors.[12]

Unfortunately for the Iranians, the tactic backfired, triggering an armed rebellion. Even worse from Tehran's perspective, while the demonstrations represented all segments of Syrian society, the civil war was sectarian, dominated by the Sunnis who were backed by Saudi Arabia and Turkey. Within months Syria, once considered a seemingly stable nation-state, came to resemble the "tribes with flags" landscape which, as noted in Chap. 1, the IRGC had normally welcomed. With Assad losing control over swaths of his country, the IRGC-QF suffered serious setbacks. For instance, Zabadani, the hub of the IRGC activity, was overrun by the Free

[11] RF, Khamenei on Arab Spring: Suppressions are not useful, *Radio Farda*, June 4, 2011, https://www.radiofarda.com/a/f2_iran_khamenei_speech_khomeini_anniversary_arab_spring_crackdown_useless/24215310.html.

[12] Afshon Ostovar, Vanguard of the Imam: religion, politics, and Iran's Revolutionary Guards (Oxford University Press, 2016).

Syrian Army (FSA)-Al-Nusra Front. The Iranian media speculated that the civil war could lead to the collapse of the Assad dynasty or the disintegration of the country into separate ethnic and sectarian enclaves.[13]

Suleimani, however, left little doubt that neither prospect was acceptable to Iran. Speaking in Qom on May 22, 2011, the QF chief emphasized that the situation in Syria was as important to the Islamic Revolution as the war with Iraq: "Today, Iran's victory or defeat would not be determined in Mehran…our borders have spread." By referring to Iraq, Suleimani could put the war in Syria within the broader context of international relations. The QF chief asserted that the United States and its allies in the Middle East had waged war against Assad to destroy the regime. As he put it, "Assad was part of a much larger Western plot to sow division in the Middle East, and weaken Iran's place in the region." Echoing the Supreme Leader, the QF chief described the struggle in Syria as a "battle between Islam and the infidels, good versus evil."[14]

With so much at stake, the IRGC-QF increased its presence in Syria. In 2011, there was reportedly 200–400 IRGC-QF personnel, providing intellectual and advisory help. At the beginning of 2012, the number reached around 1,200–2,000. They took up positions at the Damascus airport, Aleppo, Hama, and Latakia. They were backed up by some 7,000–10,000 fighters from the Lebanese Hezbollah. In one of its first decisions, the IRGC-QF moved to create the National Defense Forces (NDF) to supplement the regular Syrian army, which was depleted by the defection of the Sunni conscript and commanders. The Iranians were also not happy with the performance of the troops and felt that a less traditionally structured force would perform better in the complex war arena. Until a new force could be fielded, according to Mujahedin-e Khalq Organization (MEK) sources, the IRGC-QF sent considerable reinforcements, bringing the number of its personnel to around 8,000 to 10,000.[15]

[13] Sly, Liz. "Syria's Zabadani is liberated," but for how long? *The Washington Post*, January 1, 2012. Accessed January 10, 2014, http://www.washingtonpost.com/world/middle_east/syrias-zabadani-liberated-but-for-howlong/2012/01/21/gIQAMhDYGQ_story.html.

[14] Qassem Suleimani, Victor of Faith and Virtue over Terrorism, in Mohammad Khalilpour, A Different Story of Logic of Being in Syria (Tehran: Masir-e Aftab, 2017), pp. 31–42; Afshon Ostovar, Vanguard of the Imam: religion, politics, and Iran's Revolutionary Guards (Oxford University Press, 2016).

[15] NCRI, How Iran Fuels Syria War: Details of the IRGC Command HQ and Key Officers in Syria, *NCRI-U.S. Representative Office*, 2016.

In time, the Iranians created a substantial auxiliary force to fight alongside the Syrian army: the so-called New Syrian Force (NSF), also known as Popular Defense, and the Syrian Basij, and other militias including the NDF, The Baath Battalions, the Jerusalem Brigade, the Syrian Resistance, the Syrian Social Nationalist Party (SSNP) and the Popular Front for the Liberation of Palestine-General Command (PFLP-GC territory of the Iranian forces and their militias including the Lebanese Hezbollah, Pakistani Zeinabiouyn Brigade, the Afghani Fatemiyoun Brigade, Kataib Hezbollah and Liwa Zulfiqar Brigade).[16]

The Obama administration, as mentioned in the preceding chapter, was reluctant to call out Iran's revolutionary export activities. But the unprecedented scope of the Iranian deployment forced the president's hand. In May 2011, the U.S. Treasury sanctioned Suleimani, Chizari, and other Guards for terror and human rights abuse in Syria. One month later, on June 29, 2011, Treasury designated LEF as a terror organ for providing "material support to the Syrian General Intelligence Directorate (GID) and dispatching personnel to Damascus to assist the Syrian government in its efforts to suppress the Syrian people." The Treasury also sanctioned Ahmadi-Moghadam, the LEF chief, and his deputy Radan. Three months earlier in April 2011, Radan visited Damascus and met with Syrian security officials and provided expertise to aid in Assad's crackdown on the Syrian people.[17]

The Treasury also cited Iran's Mahan Air and Yas Air, the air cargo company, for transporting troops and cargo to Syria. To fight Assad's bad international publicity, Suleimani proposed in 2011 to create the so-called "binary option" strategy. In other words, he wanted to create a radical jihadist group that would make Assad look like a lesser evil in the eyes of his people and the West. Radical Islamist prisoners housed in special jails were promptly released. Zahran Alloush, Abu Shadi Aboud (brother of

[16] Chris Cillizza, Trump's change of heart on Syria, April 7, 2017, edition.cnn.com/2017/04/06/politics/trump-syria-foreign-policy/index.html; Rudaw, SDF enter east Raqqa in 'Great Battle' for ISIS stronghold, June 6, 2017, www.rudaw.net/english/middleeast/syria/060620171; Julian Borger, The war after Isis': has Trump opened the door to conflict with Iran?, *The Guardian*, June 19, 2017, https://www.theguardian.com/us-news/2017/jun/18/donald-trump-syria-iran-isis-james-mattis.

[17] USDT, Administration Takes Additional Steps to Hold the Government of Syria Accountable for Violent Repression Against the Syrian People, May 18, 2011, *U.S. Department of the Treasury*, https://www.treasury.gov/press-center/press-releases/Pages/tg1181.aspx.

Hassan Aboud), and Ahmed Abu Issa were among the more prominent jihadists who owned their liberty to the new strategy. Many went on to lead Islamist groups, such as Jaysh al-Islam, Ahrar ash-Sham and Suqour al-Sham Brigade, in the Syrian Civil War. When the Islamic State of Iran and Syria (ISIS) forces occupied northern Syria and proclaimed Raqqa to be the capital of its caliphate, Assad was not fazed. In fact, he helped the organization by buying ISIS oil and gas, which, according to estimates, financed 70 percent of its budget. Abu Bakr Al-Baghdadi's notorious brutality more than vindicated Suleimani's "binary option" maneuver.[18]

With the crisis deepening, the number of IRGC-QF casualties had increased as well. According to MEK, high-ranking IRGC officers, including several brigade generals suffered particularly high losses. The killing of the Brigadier General Hassan Shateri, who was close to Suleimani, on February 14, 2013, was a particularly hard blow to the Guards, who gave their fallen comrade a lavish funeral. Despite the IRGC propaganda, the public opinion, still overshadowed by the monumental losses in the Iran–Iraq war, had come to question the cost of the involvement. In 2013, Hassan Rouhani, who ran against the hardline Principalist, Saeed Jalili, raised the issue of the high cost of such an "activist" foreign policy.[19]

But Rouhani had little success in curtailing the operations in Syria. Ayatollah Mohammed Taqi Mesbah Yazdi and other Haqqani School clerics declared that Iran's involvement was part of a spiritual mandate to defend the shrine of Sayyidah Zaynab, the daughter of Imam Ali, located in Damascus. The Iran-based Grand Ayatollah Kazem Haeri, one of the mentors of Muqtada al-Sadr, went further. He decreed the fighting in Syria to be a "battle of infidels against Islam" and promised that the fallen martyrs would be rewarded in the afterlife. Not coincidentally, Suleimani, who for most of his career had operated in the shadows, had turned into something of a celebrity. Once referred to as the "ghostly commander,"

[18] Afshon Ostovar, Vanguard of the Imam: religion, politics, and Iran's Revolutionary Guards (Oxford University Press, 2016); C.J. CHIVERS, Behind the Black Flag: The Recruitment of an ISIS Killer, *The New York Times*, December 20, 2015, https://www.nytimes.com/2015/12/21/world/middleeast/isis-recruitment-killer-hassan-aboud.html; Rania Abouzeid, The Jihad Next Door, June 23, 2014, www.politico.com/magazine/story/2014/06/al-qaeda-iraq-syria-108214_full.html#.WWnoRIiGPIU.

[19] NCRI, How Iran Fuels the Syria Civil War, *National Council of Resistance of Iran*, 2016, 54–65; Entekhab, The Martyrdom of General Shateri in the Damascus-Beirut route, *Entekhab.ir*, February 14, 2013, www.entekhab.ir/fa/news/96412/شهادت-سردار-شاطری-در-مسیر-دمشق-بیروت-عکس.

he was now celebrated as the leader of the "Sacred Defense," a name previously reserved for the Iran–Iraq war.[20]

Despite the best efforts of the IRGC-QF, by 2015, Assad was losing control of the country. The combination of a robust ISIS presence around Raqqa and the unexpected vigor of a plethora of anti-Assad militias seemed to spell doom for the regime. Anti-Assad forces such as FSA backed by Turkey and the United States, Ahrar al-Sham funded by Turkey and Saudi Arabia, Al-Nusra Front backed by Saudi Arabia and Turkey, Jaysh al-Islam backed by Turkey, Qatar and Saudi Arabia, Jaish al-Fateh (the Army of Conquest) supported by Qatar and Turkey, and Syrian Democratic Forces, emerged to fight Assad.[21]

As indicated in Chap. 3, the desperate QF commander traveled to Moscow to persuade Vladimir Putin to intervene in the civil war. The Russian leader, who was looking to improve his influence in the Middle East, decided to get involved under the pretext of fighting ISIS. The Russian bombing raids targeted anti-Assad rebels in strategic strongholds such as Aleppo and Homs. Within less than a year after the Russian intervention of September 2015 turned the tide of war in Assad's favor.[22]

Enlisting Russian help was not risk-free for the Revolutionary Guards, who nursed an ambitious plan to turn Syria into an Iranian neo-protectorate. Unlike a colonial-era protectorate, Syria was to retain its sovereignty but was expected to tow the Iranian line in every other respect. This neo-protectorate was to turn Syria into a de facto 35th province of Iran to protect it from a variety of enemies. Speaking for the Guards, Hassan Abbassi made this point clear: "They say they have not come to destroy the Syrian government, but their goal is to cut Iran's hands, weaken Hezbollah, and ultimately secure Israel. Therefore, the Americans

[20] Shia News, Issuing a fatwa for Jihad in Syria by a religious authority, *Shia News*, December 5, 2013, www.shia-news.com/fa/news/64502/صدور-فتوای-وجوب-جهاد-در-سوریه-توسط-یکی-از-مراجع-دینی; Afshon Ostovar, Vanguard of the Imam: religion, politics, and Iran's Revolutionary Guards (Oxford University Press, 2016); Tyler Durden, Meet the Ghostly Iranian Spymaster, *Zero Hedge*, October 12, 2015; ZH, "Mid East Coup," *Zero Hedge*, October 1, 2015.

[21] Reuters, al-Qaeda splinter group declares Islamic 'Caliphate', *Reuters*, June 29, 2014, uk.reuters.com/article/uk-syria-crisis-iraq/al-qaeda-splinter-group-declares-islamic-caliphate-idUKKBN0F40SD20140629.

[22] Zack Beauchamp, Russia says it's bombing ISIS in Syria. This map shows it's lying, October 7, 2015, https://www.vox.com/2015/10/7/9471271/russia-syria-bombing-map.

started the war [in Syria] with us, not with Syria. The fall of Syria means that we should fight against the enemy in our own borders."[23]

Brigadier General Yadullah Javani, the head of the Guards political bureau and Ayatollah Khamenei's military adviser also noted that "today our support for the people of Iraq and Syria is an honest support… unchangeable and will not end."[24] Even those who considered the neo-protectorate to be far-fetched agreed that Syria was crucial the Shiite "axis of resistance" on several counts. For starters, the IRGC-QF utilized the civil war to build up a large paramilitary organization. To overcome the shortage of local Shiite recruits, the Guards began experimenting with creating a Shiite "foreign legion." By 2016, several militias took shape. Including the Lebanese Hezbollah in Syria, Pakistani Zainebiyoun Brigade with more than 1,000 fighters, the Afghani Abouzar Brigade under the new name of Fatemiyoun Brigade with reportedly between 12,000–20,000 fighters, Kataib Hezbollah with 400–500 fighters operating in Syria and Iraq, and the Liwa Zulfiqar Brigade claimed to have more than 7,000 fighters.[25]

In yet another variation on the Foreign Legion, the Guards created the Shiite Liberation Army (SLA), according to Mohammed Ali al-Falaki, the former commander of the 10th Imam Hossein Brigade. Al-Falaki emphasized that the new militia would destroy Israel in 23 years. The Guards recruited thousands of undocumented Afghan Shiite refugees in Iran. According to the report some of the Afghan recruits were coerced to fight in Syria out of fear of deportation. Others volunteered out of religious conviction or to receive the stipends that helped their desperately poor families. According to IRGC media, following a 30-day training session in the Guards' Imam Hossein University in Tehran, they were sent to Syria as "shrine's defense forces," a generic name given to all volunteers. According to Lieutenant General Hamid Abazari, commander of the Imam Hossein University, the rate of attrition of foreign fighters was high, notably among the Afghans who lost some 30 percent of their ranks.[26]

[23] Hassan Abbassi, Syria: Islamic World's Strategic Depth, in Mohammad Khalilpour, A Different Story of Logic of Being in Syria (Tehran: Masir-e Aftab, 2017), pp. 57–74.

[24] Yadullah Javani, Iran's Fundamental and Sincere Defending of Regional Nations, in Mohammad Khalilpour, A Different Story of Logic of Being in Syria (Tehran: Masir-e Aftab, 2017), pp. 103–109.

[25] Ali Alfoneh, Shia Afghan Fighters in Syria, *Atlantic Council*, April 19, 2017, www.atlanticcouncil.org/blogs/syriasource/shia-afghan-fighters-in-syria.

[26] Iran in Syrian War, *Twitter*, August 29, 2017, https://twitter.com/warreports/status/902508690783301633.

Having fought the last large-scale war in the 1980s, the IRGC used the Syrian theater for training and developing new technologies. Brigadier General Morteza Saffari, a senior IRGC commander at the Imam Hossein University, disclosed that cadets were deployed in Syria as part of an educational program designed for future officers. The Institute for the Study of War (ISW) noted that the IRGC, which worked closely with the Russian military, studied important Russian military concepts, such as "cauldron battles, multiple simultaneous and successive operations, and frontal aviation."[27]

Improving the aviation skills of the Guards was high on the training agenda at the Tiyas Military Air Base, located on the outskirt of Homs in central Syria. The IRGC used the base to patrol the American fleet in the eastern Mediterranean. The Niruye Daryai Sepah-e Iran (NEDSA), the naval unit of the Revolutionary Guards, was stationed at a naval facility in Latakia. To improve its drone technology, the Guards built a drone airfield near Palmyra, 80 miles from al-Tanf, at the intersection of the Iraqi, Jordanian and Syrian borders. On March 12, 2017, the media reported that Tehran was planning to establish a military base in the Syrian province of Latakia.[28]

While the use of the Hezbollah militants and the foreigner volunteers helped reduce Iran's human losses, the financial cost of the Syrian deployment was high. During his presidential campaign in 2017, Rouhani, who stressed the importance of economic recovery, criticized the IRGC for the war in Syria. He pointed out that this interventionist policy generated

[27] Ahmad Majidyar, Iran's Ground Forces Expanding Uses of Drones," *Middle East Institute, Iran Observed*, April 13, 2017; Paul Bucala and Genevieve Casagrande, How Iran is Learning from Russia in Syria, February 03, 2017, www.understandingwar.org/backgrounder/how-iran-learning-russia-syria; Ahmad Majidyar, Iran's Ground Forces Expanding Uses of Drones," *Middle East Institute, Iran Observed*, April 13, 2017; Paul Bucala and Genevieve Casagrande, How Iran is Learning from Russia in Syria, February 03, 2017, www.understandingwar.org/backgrounder/how-iran-learning-russia-syria.

[28] Psychological war of the Presence of Iran's Fighter Jets in Syria, alwaght.com/fa/News/52402/جزئیات-و-عکس--سوریه-در-ایران-های-جنگنده-حضور-روانی-جنگ; Dylan Malyasov, Iran deploy 2 dozen fighter jets on Tiyas Airbase in Syria, December 3, 2015, defence-blog.com/news/iran-deploy-2-dozen-fighter-jets-on-tiyas-airbase-in-syria.html; Iran may seek naval bases in Yemen or Syria: chief of staff, November 27, 2016, www.reuters.com/article/us-iran-navy-yemen-syria-idUSKBN13M08M; TAAS, Iran's Syrian base gets green light, *TAAS News Agency*, March 13, 2017, tass.com/pressreview/935105; Courtney Kube, Iran Has Drone Base Near U.S. Troops in Syria, June 20, 2017, www.nbcnews.com/news/world/iran-has-drone-base-near-u-s-troops-syria-n774811; Iran's Agreement with Bashar Assad for Building Military Base in Latikka, https://tnews.ir/news/793b81868435.html.

negative publicity and triggered sanctions, which the country, still recovering from the nuclear sanctions, could not afford. The former reformist mayor of Tehran Gholam Hossein Karbaschi publicly questioned the cost of war in Syria and said, "We too want peace…for the oppressed to be defended and the Shiites to be supported. But this cannot be done by giving money, buying arms and killing." In a rare display of public discontent, students at the Azarbaijan Shahid Madani University denounced Hassan Abbassi for his role in providing an intellectual justification for what they considered to be an ill-advised adventure.[29]

The Guards and their surrogates responded with a public relations campaign of their own. They took groups of journalists on tours of Syria to help explain why their critics were wrong. Media associated with the IRGC reminded its readers that Syria was "another sacred defense" while warning that for the United States and the "Zionists" destroying the shrines in Syria was a prelude to the destruction of the Shiite shrines in Iran. Hardline media took to carrying images of the Supreme Leader comforting families of "fallen martyrs."[30]

On one such occasion on June 25, 2016, Khamenei stated: "I have always said that our martyrs are on the front lines defending Islam and that their families—including fathers, mothers, wives and children—stand, immediately, behind them; this should be appreciated." He added that "the issue, with regard to families of martyrs, is not that they have merely lost a martyr on the path of truth; their patience has mountains of value. The fact that a martyr's father, mother, spouse, or children endure the loss of their dear one is a lofty value."[31]

Rouhani's decisive victory in the election against Ibrahim Raisi, which favored the Supreme Leader and the Guards, evidently rattled the Principalists. So much so that Khamenei felt compelled to extol the virtue of martyrdom in the service of revolutionary export. On June 18, 2017, Khamenei gave a major address to the martyred families: "Today,

[29] Ahmad Majidar, Iran Observed, *Middle East Institute*, May 15, 2017; Gholamhossein Karbasch Criticized Iran's Syria Policy, https://www.youtube.com/watch?v=tU8MECPO7ZY&t=26s.

[30] Amir Taheri, Syria, Iranians Find it Not So Simple," *Asharq AlAwsat*, May 19, 2017, https://english.aawsat.com/amir-taheri/opinion/syria-iranians-find-not-simple.

[31] "Khamenei.ir, Families of martyrs are also on the front lines of defending Islam August 16, 2017, english.khamenei.ir/news/5063/Families-of-martyrs-are-also-on-the-front-lines-of-defending.

there is a war which our people do not feel with their flesh and blood, but at the same time, they participate. They go to the war, even though, they do not feel it at home, why? Because they understand the situation. This is the miracle of the revolution." He added that "the main reasons for the endurance of the Islamic Revolution and its might are great concepts and values such as 'martyrdom, jihad, patience and perseverance of the families of martyrs and the enemy tries to dispossess the society of these values and unfortunately some people participate in this framework."[32]

Given the negotiated political order, Khamenei need not have bothered drumming up public support. Much as Rouhani's success at the polls indicated popular dissatisfaction with the adventurism of the IRGC-QF, the president had little input into the parastatal domain. The terror attacks on the Majlis on June 7, 2017, carried out by a jihadist Kurdish group made matters worse for normalizers. The hardliners blamed the Rouhani government for the attack. Kayhan newspaper wrote: "When you [Rouhani] slammed and weakened the Revolutionary Guards for testing missiles, when you questioned all of the past 38 years, it gave a good opportunity to America, Israel, and Saudi Arabia to fill the gap with security challenges." Others noted that Rouhani was premature when he disparaged the services of the Revolutionary Guards as the protectors of the republic.[33]

Hamid Rasaei, who represented the Principalists under former President Mohammed Ahmadinejad stated, "Terrorists are here to help coward statesmen." Others went further, calling Rouhani a traitor and sarcastically asking: "Wasn't it you that just said there would be war if I am not elected?" Hardliners also attacked Karbaschi and other critics of the IRGC's commitment to Assad. Ahmad Tavakoli, a conservative politician, and journalist, wrote on his Instagram: "When I heard about the

[32] Khamenei.ir, Families of martyrs are also on the front lines of defending Islam August 16, 2017, english.khamenei.ir/news/5063/Families-of-martyrs-are-also-on-the-front-lines-of-defending; Islamic Republic firmly standing and nation to slap enemy in face, www.leader.ir/en/content/18553/The-Leader-meeting-with-families-of-Defenders-of-Shrine-and-Border-Guard-martyrs; Khamenei.ir, Speech in Visiting of families of Shrine Defenders Martyrs, farsi.khamenei.ir/speech-content?id=32,186.

[33] Kayhan Newspaper, Terrorist Attacks is a Result of Giving Wrong Address to America and Israel, *Kayhan Newspaper*, June 8, 2017, kayhan.ir/fa/news/105821.

ISIS attack…I told myself 'congratulations' to Mr. Karbaschi and his like-minded ones."[34,35]

While the Guards had little to fear from Rouhani, the election of Donald Trump cast doubts on the neo-protectorate plan for Syria.

From a Neo-Protectorate to a Syrian Federation?

To recall Chap. 2, the administration, acting in tandem with Congress, increased pressure on Iran, but IRGC's presence in Syria escaped scrutiny because President Donald Trump considered it secondary to his agenda. The administration changed its mind, however, after the regime's air force dropped gas munitions on civilians in northern Idlib province on April 4, 2017. The attack killed 70 people including 20 children. Two days later, the U.S. navy used several Tomahawk missiles to hit the airfield where the airplanes took off. In a special statement, the White House explained that "years of previous attempts at changing Assad's behavior have all failed, and failed very dramatically." The statement stopped short of calling for the toppling of Assad but signaled a change from Obama's noninterventionist policy.[36]

While commentators attributed Trump's change of heart to the images of carnage, the Department of Defense was worried about Iran's neo-protectorate scheme. Not coincidentally, the Pentagon was drawing up plans to liberate Raqqa and the adjacent territory to the north. Having collaborated with the Syrian Kurdish forces, the military began an intensive program of training other anti-Assad groups, including Fursan al-Haq, or Knights of Righteousness, and the Kurdish-dominated Syrian Democratic Forces (SDF). The IRGC-QF, still hoping to create the neo-protectorate, was highly anxious about the new development,

[34] Karbaschi Criticizing Iran's Policies in Syria, *Iranian TV Channel*, April 29, 2017, https://www.youtube.com/watch?v=tU8MECPO7ZY; Shargh Newspaper, Political Fishing out of Terror Attacks, *Shargh Newspaper*, June 8, 2017, www.sharghdaily.ir/News/133090/ماهی‌گیری-سیاسی-از-حمله-تروریستی.

[35] Rouhani: Iran helped Iraq, Syria and Lebanon and will Continue it, iranintl.com/سیاست/روحانی-ایران-به-عراق،-سوریه-و-لبنان-کمک-کرده-و-خواهد-کرد.

[36] Martin Chulov and Kareem Shaheen, Syria chemical weapons attack toll rises to 70 as Russian narrative is dismissed, April 5, 2017, https://www.theguardian.com/world/2017/apr/04/syria-chemical-attack-idlib-province; WH, Statement of the President on Syria, *White House*, https://www.whitehouse.gov/the-press-office/2017/04/06/statement-president-trump-syria.

not least because of the American support for an autonomous Syrian–Kurdish territory. Still worse, the Guards commanders realized that the American and their proxies wanted to disrupt the planned land bridge, which was discussed in the previous chapter.

According to Kurdish intelligence sources, in anticipation of such a complication, the IRGC moved the original Syrian section of the route 150 miles to the south. The new idea was to secure a land bridge from Iran to Lebanon. The primary path was driven from Baquba, Shirqat and Rabia (Tal Afar) in Iraq to Sinjar, Qamishli, Kobani, Idlib, Homs and Latika in Syria. However, in order to avoid American-supported forces marching on the Islamic State group stronghold of Raqqa, Iran diverted the course of a land corridor. According to Kurdish intelligence sources, the IRGC moved the original Syrian section of the course 150 miles to the south. The new line would traverse a distance of at least 800 miles from Iran's western borders and is expected to pass through Mayadin, Deir al-Zour, to Sukhna to Palmyra, then Damascus and Homs, and end at the port of Latakia. The pathway provided a link to Hezbollah in Lebanon, finally ending at the edge of the Golan Heights. The corridor would serve as chain to move military supplies or militiamen when needed.[37]

Even though the Guards were careful to draw the new route in accordance with the de-conflicting zones of the Astana agreement described in Chap. 3, the United States, which was not a party to the talks, had different ideas. By carefully seizing the strategic stronghold of ISIS as well as others, the American military tried to disrupt the land bridge. Tanf, where an American garrison trained anti-Assad fighters, was key to this effort. The Americans declared the area around Tanf a "deconfliction zone"; that is, a 55-kilometer exclusion zone. When Suleimani challenged the no-go zone on June 6, by sending one of the local militias, American warplanes bombed the advancing contingent. On June 8, 2018 an American jet was shot down by an IRGC Shahed-129 drone flying in the exclusion zone.[38]

[37] Baghdad Post, Why Iran is Building Two Land Corridors to the Mediterranean, May 01, 2017, www.thebaghdadpost.com/en/story/10029/Why-Iran-is-Building-Two-Land-Corridors-to-the-Mediterranean; Baghdad Post, Revealed: Iran's Shiite corridor to Mediterranean Sea via Tanf area, May 23, 2017, www.thebaghdadpost.com/en/story/11003/Revealed-Iran-s-Shiite-corridor-to-Mediterranean-Sea-via-Tanf-area.

[38] Fox News, U.S. air strikes pound pro-Assad forces in Syria, *Fox News*, May 18, 2017, www.foxnews.com/world/2017/05/18/us-air-strikes-pound-pro-assad-forces-in-syria.html; Carla Babb, Mattis: Pro-Syrian Government Forces in Deconfliction Zone Were 'Iranian-directed', May 19, 2017, https://www.voanews.com/a/jim-mattis-pro-syrian-government-

And on June 18, a U.S. Air Force F/A-18E Super Hornet shot down a Syrian plane, which allegedly bombed the Syrian Democratic Forces, an American ally. Joseph Dunford, chairman of the Joint Chiefs of Staff, explained that the action was justified because the Americans needed to protect their personnel. As he put it, this "was a force protection strike. Our commanders on the ground felt like they were threatened at that point. And their rules of engagement allow them to do that."[39]

Even so, Iranian government officials were careful to limit their reaction to verbal condemnations. Foreign Ministry spokesman Bahram Qasemi stated that "Iran strongly condemns any such unilateral strikes…such measures will strengthen terrorists in Syria…and it will complicate the situation in Syria and the region." Iran, along with Russia, have warn that they will "respond with force" if their own "red lines" are crossed in Syria. The two allies joint statement read, "America waged an aggression on Syria and crossed the red lines. From now on we will respond with force to any aggressor or any breach of red lines from whoever it is and America knows our ability to respond well," the group's joint command center said.[40]

Interestingly, the IRGC-QF kept a relatively low profile, at least compared to its previous threats. For example, in August 2013, Suleimani sent a "fire and brimstone" message to the White House via Nuri al-Maleki. He declared that "any parachute that is going to be hell-born, each soldier that is going to get off from the navy carrier, each commando that is going to be deployed, should make sure that he has already ordered his coffin.

forces-deconflication-zone-iranian-directed/3862614.html; Michael R. Gordon, June 20, 2017, American Warplane Shoots Down Iranian-Made Drone Over Syria, *The New York Times*, https://www.nytimes.com/2017/06/20/world/middleeast/american-warplane-shoots-down-iranian-made-drone-over-syria.html.

[39] LaTimes, U.S. warplane shoots down Syrian jet near Raqqah, *Los Angeles Times*, June 18, 2017, www.latimes.com/world/middleeast/la-fg-syria-us-jet-20170619-story.html; USDoD, Department of Defense Press Briefing by Secretary Mattis, General Dunford and Special Envoy McGurk on the Campaign to Defeat ISIS in the Pentagon Press Briefing Room, *Defense.gov*, May 19, 2017, https://www.defense.gov/News/Transcripts/Transcript-View/Article/1188225/department-of-defense-press-briefing-by-secretary-mattis-general-dunford-and-sp/; Dmitry Lee and Galiaskar Seitzhan, Latest Astana process talks establish de-escalation zones in war-torn Syria, May 7, 2017, astanatimes.com/2017/05/latest-astana-process-talks-establish-de-escalation-zones-in-war-torn-syria/.

[40] Ali Kushki U.S. draws Iran rebuke over missile attack on Syrian airfield, April 7, 2017, www.tehrantimes.com/news/412404/U-S-draws-Iran-rebuke-over-missile-attack-on-Syrian-airfield; Russia and Iran have warned the U.S. they will "respond with force" if their own "red lines" are crossed in Syria, *The Independent*, April 9, 2017, www.independent.co.uk/news/world/middle-east/russia-iran-us-america-syria-red-lines-respond-with-force-aggressor-air-strikes-war-latest-a7675031.html.

Afshon Ostovar, Vanguard of the Imam: religion, politics, and Iran's Revolutionary Guards (Oxford University Press, 2016).

Ahmad Majidar, Iran Observed, *Middle East Institute*, May 15, 2017.

Ahmad Majidyar, "Iran's Ground Forces Expanding Uses of Drones," *Middle East Institute, Iran Observed*, April 13, 2017.

Ali Akbar Velayati, Syria is the Golden Ring of Resistance line, in Mohammad Khalilpour, A Different Story of Logic of Being in Syria (Tehran: Masir-e Aftab, 2017).

Ali Alfone, Iran Unveiled: How the Revolutionary Guards Is Transforming Iran from Theocracy into Military Dictatorship (AEI Press, 2013).

Ali Alfoneh, Shia Afghan Fighters in Syria, *Atlantic Council*, April 19, 2017, www.atlanticcouncil.org/blogs/syriasource/shia-afghan-fighters-in-syria.

Ali Kushki U.S. draws Iran rebuke over missile attack on Syrian airfield, April 7, 2017, www.tehrantimes.com/news/412404/U-S-draws-Iran-rebuke-over-missile-attack-on-Syrian-airfield.

Amir Taheri, "Syria, Iranians Find it Not So Simple," *Asharq AlAwsat*, May 19, 2017, https://english.aawsat.com/amir-taheri/opinion/syria-iranians-find-not-simple.

Asriran, Chief of Amar Base, Syria is the 35th province of Iran, *Asre Iran*, February 14, 2013, http://www.asriran.com/fa/news/257730/ رئیس-قرارگاه-عمار-سوریه-سی-و-پنجم-است-اگر-دشمن-بخواهد-سوریه-یا-خوزستان-را-بگیرد-اولویت-حفظ-سوریه-است-تحریم-ها-مثل-قبل-نیست.

Baghdad Post, Revealed: Iran's Shiite corridor to Mediterranean Sea via Tanf area, May 23, 2017, www.thebaghdadpost.com/en/story/11003/Revealed-Iran-s-Shiite-corridor-to-Mediterranean-Sea-via-Tanf-area.

Baghdad Post, Why Iran is Building Two Land Corridors to the Mediterranean, May 01, 2017, www.thebaghdadpost.com/en/story/10029/Why-Iran-is-Building-Two-Land-Corridors-to-the-Mediterranean.

C.J. Chivers, Behind the Black Flag: The Recruitment of an ISIS Killer, *The New York Times*, December 20, 2015, https://www.nytimes.com/2015/12/21/world/middleeast/isis-recruitment-killer-hassan-aboud.html.

Carla Babb, Mattis: Pro-Syrian Government Forces in Deconfliction Zone Were 'Iranian-directed', May 19, 2017, https://www.voanews.com/a/jim-mattis-pro-syrian-government-forces-deconflication-zone-iranian-directed/3862614.html.

Chris Cillizza, Trump's change of heart on Syria, April 7, 2017, edition.cnn.com/2017/04/06/politics/trump-syria-foreign-policy/index.html.

Courtney Kube, Iran Has Drone Base Near U.S. Troops in Syria, June 20, 2017, www.nbcnews.com/news/world/iran-has-drone-base-near-u-s-troops-syria-n774811.

David M. Smick, Iran's Ahmadinejad: Crazy or Crazy Like a Fox? *The International Economy* Winter 2006.

Dmitry Lee and Galiaskar Seitzhan, Latest Astana process talks establish de-escalation zones in war-torn Syria, May 7, 2017, astanatimes.com/2017/05/latest-astana-process-talks-establish-de-escalation-zones-in-war-torn-syria/.

Dylan Malyasov, Iran deploy 2 dozen fighter jets on Tiyas Airbase in Syria, December 3, 2015, defence-blog.com/news/iran-deploy-2-dozen-fighter-jets-on-tiyas-airbase-in-syria.html.

Entekhab, General Jafari: Bashar Assad is not the Reason for War in Syria, its Resistance and Islamic Revolution, *Entekhab*, September 27, 2017, www.entekhab.ir/fa/news/367339/سردار-جعفری-هدف-جنگ-در-سوریه-بشار-اسد-نیست-مقاومت-و-انقلاب-اسلامی-است.

Entekhab, The Martyrdom of General Shateri in the Damascus-Beirut route, *Entekhab.ir*, February 14, 2013, www.entekhab.ir/fa/news/96412/شهادت-سردار-شاطری-در-مسیر-دمشق-بیروت-عکس.

Fox News, U.S. air strikes pound pro-Assad forces in Syria, *Fox News*, May 18, 2017, www.foxnews.com/world/2017/05/18/us-air-strikes-pound-pro-assad-forces-in-syria.html.

Gholamhossein Karbasch Criticized Iran's Syria Policy, *YouTube*, https://www.youtube.com/watch?v=tU8MECPO7ZY&t=26s.

Hassan Abbassi, Syria: Islamic World's Strategic Depth, in Mohammad Khalilpour, A Different Story of Logic of Being in Syria (Tehran: Masir-e Aftab, 2017).

Iran in Syrian War, *Twitter*, August 29, 2017, https://twitter.com/warreports/status/902508690783301633.

Iran may seek naval bases in Yemen or Syria: chief of staff, November 27, 2016, www.reuters.com/article/us-iran-navy-yemen-syria-idUSKBN13M08M.

Iran's Agreement with Bashar Assad for Building Military Base in Latikka, https://tnews.ir/news/793b81868435.html.

Islamic Republic firmly standing and nation to slap enemy in face, www.leader.ir/en/content/18553/The-Leader-meeting-with-families-of-Defenders-of-Shrine-and-Border-Guard-martyrs.

Julian Borger, The war after Isis': has Trump opened the door to conflict with Iran?, *The Guardian*, June 19, 2017, https://www.theguardian.com/us-news/2017/jun/18/donald-trump-syria-iran-isis-james-mattis.

Karbaschi Critisizing Iran's Policies in Syria, *Iranian TV Channel*, April 29, 2017, https://www.youtube.com/watch?v=tU8MECPO7ZY.

Kayhan Newspaper, Terrorist Attacks is a Result of Giving Wrong Address to America and Israel, *Kayhan Newspaper*, June 8, 2017, kayhan.ir/fa/news/105821.

Khamenei.ir, Families of martyrs are also on the front lines of defending Islam August 16, 2017, english.khamenei.ir/news/5063/Families-of-martyrs-are-also-on-the-front-lines-of-defending.

Khamenei.ir, Speech in Visiting of families of Shrine Defenders Martyrs, farsi.khamenei.ir/speech-content?id=32186.

LaTimes, U.S. warplane shoots down Syrian jet near Raqqah, *Los Angeles Times*, June 18, 2017, www.latimes.com/world/middleeast/la-fg-syria-us-jet-20170619-story.html.

Martin Chulov and Kareem Shaheen, Syria chemical weapons attack toll rises to 70 as Russian narrative is dismissed, April 5, 2017, https://www.theguardian.com/world/2017/apr/04/syria-chemical-attack-idlib-province.

Michael R. Gordon, June 20, 2017, American Warplane Shoots Down Iranian-Made Drone Over Syria, *The New York Times*, https://www.nytimes.com/2017/06/20/world/middleeast/american-warplane-shoots-down-iranian-made-drone-over-syria.html.

MN, A Message from Qassem to America about Syria, *Mashregh News*, https://www.mashreghnews.ir/news/245283/سوریه-درباره-آمریکا-به-قاسم-حاج-پیام.

NCRI, How Iran Fuels Syria War: Details of the IRGC Command HQ and Key Officers in Syria, *NCRI-U.S. Representative Office*, 2016.

NCRI, How Iran Fuels the Syria Civil War, *National Council of Resistance of Iran*, 2016.

Paul Bucala and Genevieve Casagrande, How Iran is Learning from Russia in Syria, February 03, 2017, www.understandingwar.org/backgrounder/how-iran-learning-russia-syria.

Payvand, Defense Minister: Iran, Syria defense cooperation at highest level, June 16, 2006, www.payvand.com/news/08/may/1278.html.

Psychological war of the Presence of Iran's Fighter Jets in Syria, alwaght.com/fa/News/52402/جزئیات-و-عکس--سوریه-در-ایران-های-جنگنده-حضور-روانی-جنگ.

Qassem Suleimani, Victor of Faith and Virtue over Terrorism, in Mohammad Khalilpour, A Different Story of Logic of Being in Syria (Tehran: Masir-e Aftab, 2017).

Rania Abouzeid, The Jihad Next Door, June 23, 2014, www.politico.com/magazine/story/2014/06/al-qaeda-iraq-syria-108214_full.html#.WWnoRIiGPIU.

Reuters, al-Qaeda splinter group declares Islamic 'Caliphate', *Reuters*, June 29, 2014, uk.reuters.com/article/uk-syria-crisis-iraq/al-qaeda-splinter-group-declares-islamic-caliphate-idUKKBN0F40SD20140629.

RF, Khamenei on Arab Spring: Suppressions are not useful, *Radio Farda*, June 4, 2011, https://www.radiofarda.com/a/f2_iran_khamenei_speech_khomeini_anniversary_arab_spring_crackdown_useless/24215310.html.

Rouhani: Iran helped Iraq, Syria and Lebanon and will Continue it, iranintl.com/کرد-خواهد-و-کرده-کمک-لبنان-و-سوریه-عراق،-به-ایران-روحانی/سیاست.

Rudaw, SDF enter east Raqqa in 'Great Battle' for ISIS stronghold, June 6, 2017, www.rudaw.net/english/middleeast/syria/060620171.

Russia and Iran have warned the U.S. they will "respond with force" if their own "red lines" are crossed in Syria, *The Independent*, April 9, 2017, www.independent.co.uk/news/world/middle-east/russia-iran-us-america-syria-red-lines-respond-with-force-aggressor-air-strikes-war-latest-a7675031.html.

Shargh Newspaper, Political Fishing out of Terror Attacks, *Shargh Newspaper*, June 8, 2017, www.sharghdaily.ir/News/133090/تروریستی-حمله-از-سیاسی-ماهیگیری.

Shia News, Issuing a fatwa for Jihad in Syria by a religious authority, *Shia News*, December 5, 2013, www.shia-news.com/fa/news/64502/دینی-مراجع-از-یکی-توسط-سوریه-در-جهاد-وجوب-فتوای-صدور.

Sly, Liz. "Syria's Zabadani is liberated," but for how long? *The Washington Post*, January 1, 2012, Accessed January 10, 2014, http://www.washingtonpost.

com/world/middle_east/syrias-zabadani-liberated-but-for-how-long/2012/01/21/gIQAMhDYGQ_story.html.

TAAS, Iran's Syrian base gets green light, *TAAS News Agency*, March 13, 2017, tass.com/pressreview/935105.

The Result of Iran's President's Visit to Syria, January 20, 2006, www.nourizadeh.com/archives/001640.php#more.

Tyler Durden, Meet the Ghostly Iranian Spymaster, *Zero Hedge*, October 12, 2015.

USDoD, Department of Defense Press Briefing by Secretary Mattis, General Dunford and Special Envoy McGurk on the Campaign to Defeat ISIS in the Pentagon Press Briefing Room, *Defense.gov*, May 19, 2017, https://www.defense.gov/News/Transcripts/Transcript-View/Article/1188225/department-of-defense-press-briefing-by-secretary-mattis-general-dunford-and-sp/.

USDT, Administration Takes Additional Steps to Hold the Government of Syria Accountable for Violent Repression Against the Syrian People, May 18, 2011, *U.S. Department of the Treasury*, https://www.treasury.gov/press-center/press-releases/Pages/tg1181.aspx.

WH, Statement of the President on Syria, *White House*, https://www.whitehouse.gov/the-press-office/2017/04/06/statement-president-trump-syria.

Yadullah Javani, Iran's Fundamental and Sincere Defending of Regional Nations, in Mohammad Khalilpour, A Different Story of Logic of Being in Syria (Tehran: Masir-e Aftab, 2017).

Zack Beauchamp, Russia says it's bombing ISIS in Syria. This map shows it's lying, October 7, 2015, https://www.vox.com/2015/10/7/9471271/russia-syria-bombing-map.

ZH, "Mid East Coup," *Zero Hedge*, October 1, 2015.

CHAPTER 7

Iran and Saudi Arabia: The Struggle for Regional Hegemony and Islamic Primacy

In a region long dominated by colonial interests, Iran and Saudi Arabia found themselves on the same side of the divide. Guided by neorealism, both countries aligned themselves to the United States, a decision that earned them the role of the "twin pillars" of the Gulf. While occasional tension intruded, the royal houses of Iran and Saudi Arabia had cordial relations, which Washington had carefully nurtured.

There was nothing in the history of the two monarchies to predict the wrenching change that the Islamic Revolution had ushered. Although the pragmatic neorealists in the new regime, such as Ayatollah Mohammed Beheshti and Akbar Hashemi Rafsanjani, counseled against aggravating Riyadh, they could not change Ayatollah Ruhollah Khomeini's animosity. To the ancient Sunni–Shiite rift the Supreme Leader added some contemporary grievances. To him, Saudi Arabia was part of the so-called "American Islam," a corrupt form of Islamic subjugation to the imperialist West led by the United States. In line with the idealistic vision of international relations, Khomeini felt that the revolution export should challenge the seat of Wahhabism and unite the entire Muslim Ummah under the right Islamic creed.[1]

[1] Vali Nasr; Tali Rachel Grumet, New Middle East Cold War: Saudi Arabia and Iran's Rivalry (California: University of Denver, 2015).

F. Rezaei, *Iran's Foreign Policy After the Nuclear Agreement*, Middle East Today, https://doi.org/10.1007/978-3-319-76789-5_7

Having been challenged as the custodian of the Islamic world and anxious about the political and social upheaval among its Shiite minority, Saudi Arabia unleashed a countercampaign. The Saudis sided with Saddam Hussein during the Iran–Iraq war and manipulated the price of oil to undermine Iran's economy. Utilizing its dominant position within OPEC, Riyadh increased its oil production, a decision that the regime bitterly resented. Even the neorealists were ready to concede that the Saudis were acting on behalf of the United States, which supported Iraq. At the end of the costly war, Iran's attitude to Saudi Arabia went from the idealistic challenge of Khomenism to a full-fledged power rivalry which observers dubbed the Islamic "Cold War." The Gulf Cooperation Council (GCC) comprised of the Arab Sunni Gulf states (including Saudi Arabia, Kuwait, Bahrain, Qatar, United Arab Emirates and Oman) coordinated military and economic policy to contain Iran's reach. GCC's close relations with the United States incensed Khomeini who accused Saudi Arabia of being an agent of American imperialism.[2]

The relationship between the two countries changed for the better after the death of Ayatollah Khomeini. As the pragmatic Rafsanjani explained, "One of the wrong things we did in the revolutionary atmosphere was to make enemies. We pushed those who could be neutral into hostility and did not do anything to attract those who could become friends. It is part of the new plan that in foreign policy we should behave in a way not to needlessly leave ground to the enemy."[3]

Despite some misgivings on both sides, Iran and Saudi Arabia restored diplomatic relations in March 1991. President Khatami, known mostly for his Western-oriented Dialogue Among Civilizations, was also a keen advocate of improving relations with the kingdom. In May 1998, the two countries signed the Comprehensive Cooperation Agreement. A year later, Khatami visited Saudi Arabia to discuss issues of mutual interest. In 2001, Tehran and Riyadh signed a security cooperation agreement.

If the agreement was the high point in the détente between Tehran and Riyadh, it was also a short-lived one. Two years later, after the American invasion of Iraq, Ayatollah Ali Khamenei Iran's Supreme Leader and the revolutionary exporters seized on the opportunity to renew their campaign

[2] Chang-Cheng Liu, Saudi-Iranian Relations, 1977–1997, Durham University] Tali Rachel Grumet, New Middle East Cold War: Saudi Arabia and Iran's Rivalry (California: University of Denver, 2015), p. 64.

[3] Youssef M. Ibrahim, Iran Calls U.S. Action A 'Barbaric Massacre', *The New York Times*, July 4, 1988, www.nytimes.com/1988/07/04/world/iran-calls-us-action-a-barbaric-massacre.html.

for Shiite dominance. As noted, the Supreme Leader considered the situation God-given and, in his zeal to further the revolutionary goal, sided with the Islamic Revolutionary Guards Corps (IRGC) and Quds Force (QF) against the president. Khatami was not privy to the plans of the Revolutionary Guards to fund the Shiite militias. Within a few short years, the Revolutionary Guards had created a formidable Shiite militia force in Iraq and had penetrated the political system, as described in Chap. 5.[4]

Unsurprisingly, the growing Iranian dominance in Iraq had upset the Saudis, who refused to recognize the new Shiite-dominated government of Nouri al-Maliki. Sectarianism, always a background issue in the balance of power politics in the Gulf, took front stage. As one observer put it, "the change from a Sunni-dominated government to a...[Shiite one] shifted Baghdad from Riyadh's orbit into Tehran's...much to Saudi Arabia's discontent." Apart from its symbolism, the shift bore considerable tangible benefits to Iran in the form of lucrative oil development deals and commercial ties.[5]

Encouraged by their success in Iraq, the IRGC-QF branched further afield to Yemen, the home of the Shiite-Zaidi minority. Low on the list of revolutionary exporters, the Houthis, hailing from the Shiite tribesmen of North Yemen, were traditionally moderate and accepted Sunni Islam. After the invasion of Iraq, however, a splinter group of the movement became radicalized, engaging in large demonstrations against the United States and Israel, which alarmed the government of Ali Abdullah Saleh. Saudi Arabia, which had its Shiite minority in the Al-Hasa and Qatif provinces, was also worried since the intelligence services determined that the QF was involved in the agitation. In June 2004, a dissident cleric Hussein Badreddin al-Houthi, a leader of the Zaidi sect and a member of the Al-Haqq ("The Truth) political party started a military rebellion against the Saleh regime. When he was killed shortly after, on September 10, his brothers Abdul-Malik, Yahia, and Abdul-Karim took over. Riyadh was greatly alarmed when, in November 2009, the Houthis crossed the Saudi border and occupied several Shiite villages. Saudi Arabia responded with a full-fledged attack on the Houthi forces, a decision which prompted the IRGC to help the rebels. But facing overwhelming odds, in February 2010, the Houthis withdrew from the Saudi territory and signed a cease-fire agreement with Riyadh.

[4] Tali Rachel Grumet, New Middle East Cold War: Saudi Arabia and Iran's Rivalry (California: University of Denver, 2015), 89.

[5] Kelsey Hampton, "Doctrinally and Politically Opposed On the Battlefield in Syria: Iran and Saudi Arabia's Cold War in the Middle East" (tenth Biennial Graduate Student Conference-Conflict Studies and Global Governance: The New Generation, Boston, MA, October 31-November 1, 2014), 6.

The Arab Spring in the Gulf: From Popular Protest to Civil War

While Ayatollah Khamenei offered flowery rhetoric about the second stage of the Islamist revolution, the IRGC-QF had concrete operational plans for the Gulf region. In fact, the Guards had high hopes for Bahrain, the Shiite population of Saudi Arabia, and Yemen.

Of the three, Bahrain was of special importance. Although considered a historical province of Iran, the Shah decided to respect the referendum which the British conducted before granting the island its independence in 1971. But Ayatollah Khomeini, having described the Shiites in Bahrain as victims of Sunni oppression, revived the irredentist claim to Bahrain. Hassan Rouhani, then a Majlis member and close to the Supreme Leader, stated that "he did not agree with the policies of the Shah's regime regarding Bahrain and that the Island still belonged to Iran." He promised to "lead a revolutionary movement for the annexation of Bahrain unless its rulers adopted an Islamic form of government like the one established in Iran." Mindful of the complications with Saudi Arabia, Akbar Hashemi Rafsanjani quickly squashed the idea and the Foreign Minister Ebrahim Yazdi explained that the revolutionary chaos was responsible for such "unauthorized statements."[6]

When demonstration in Manama, the capital of Bahrain, erupted during the Arab Spring in 2011, the Revolutionary Guards seized upon the opportunity to stir up irredentist sentiment. Brigadier General Yadollah Javani, the head of the Revolutionary Guards political bureau, resurrected the long-forgotten slogans of the Office of Liberation Movement (OLM) and promised to liberate the "oppressed Shiites." Behind the scenes, the QF was working closely with several Shiite cells such as al-Wifag and the National Democratic Action Society (Waad). As one analyst put it, the IRGC-QF could count on the "Bahrain identity entrepreneurs" such as al-Wigaf.[7]

Unfortunately for the Guards, Saudi Arabia, which became greatly alarmed about the Iranian outpost on an island reached by a 25-km causeway, decided to intervene. In Operation Peninsula Shield, a sizable GCC force crushed the Shiite rebellion and left Manama and the Shiite villages

[6] Chang-Cheng Liu, Saudi-Iranian Relations, 1977–1997, Durham University, 241.

[7] Tali Rachel Grumet, New Middle East Cold War: Saudi Arabia and Iran's Rivalry (California: University of Denver, 2015), 120–124.

with "checkpoints, bunkers, and military encampments." The Saudi move outraged the Iranian hardliners who vented their frustration in public. Student groups affiliated with the Basij militia were particularly incensed; they urged the government to expel the ambassadors of Bahrain and Saudi Arabia and volunteered to go to Bahrain and "don the robes of the martyrs." The IRGC paper *Javan* fanned the flames by comparing the events to the heroic battle of Karbala, where Imam Hossein fought the overwhelming Sunni Umayyad force and perished.[8]

The Shiites fared no better in the Eastern Province of Saudi Arabia, the home of the oil and chemical industry. Compared to other Arab countries, the protest in the Kingdom was small, but it worried Riyadh because it was virtually exclusively sectarian. Sheikh Nimr Baqir al-Nimr, commonly referred to as Sheikh Nimr, a Shiite clergy with broad support in the Eastern province, challenged Riyadh. In 2009, Sheikh Nimr called for the Eastern Province's Qatif and Al-Ihsaa governorates to be separated from Saudi Arabia and united with Bahrain. King Abdullah responded by offering economic concessions, including the building of half a million housing units for the poorest segments of the population. Sheikh Nimr, however, was not placated. He urged to abolish of the House of Saud and to equalize the status for the Shiite community. As his prominence grew, Sheikh Nimr took to predicting that the House of Saud would be overthrown. For its part, Riyadh accused Tehran of manipulating the cleric and his followers. In July 2012, the government arrested Sheikh Nimr and two years later, on October 15, 2014, he was sentenced to death for "seeking foreign meddling in the Kingdom" and inciting violence.[9]

The ruling prompted a large protest in the Eastern Province. But it was the Iranians who reacted most harshly. The head of Iran's regular army, the Artesh, warned Saudi Arabia that it would "pay dearly" if the execution were carried out. Mohammad Ali Jafari, the IRGC chief, called the Saudi moves a "foolishness" which would lead to the downfall of the House of Saud. He also stated that Iran would respond to this provocation in due time: "One must have revolutionary patience…one must wait for the hand of divine revenge.

[8] Afshon Ostovar, Vanguard of the Imam Religion, Politics, and Iran's Revolutionary Guards (London: Oxford University Press, 2015), 4380.

[9] Louisa Loveluck, Nimr al-Nimr: Saudi Arabia's outspoken cleric, *Telegraph*, January 2, 2016, www.telegraph.co.uk/news/worldnews/middleeast/saudiarabia/12078497/Nimr-al-Nimr-Saudi-Arabias-outspoken-cleric.html; BBC, Saudi Shia cleric Nimr al-Nimr 'sentenced to death', *BBC News*, October 15, 2014, www.bbc.com/news/world-middle-east-29627766.

God willing, to avenge in the future the crimes they have committed." QF worked on a scheme to hit Saudi targets abroad. In May 2011, QF operatives assassinated a Saudi consular official in Karachi. But a plot to assassinate the Saudi ambassador to Washington was uncovered by the U.S. Federal Bureau of Investigation (FBI) before the QF could claim revenge.[10]

If clashing directly with Saudi Arabia was not an option, the Arab Spring in Yemen presented a perfect opportunity. In January 2011, peaceful demonstrations in Sana against the government of President Ali Abdullah Saleh had turned into sectarian riots, which brought the Houthis back into action. Following a strategy developed in Iraq and Syria, the QF seized on the opportunity to insert itself more decisively in Yemen.

Rouhani's Short Détente with Saudi Arabia: The War in Yemen as a Spoiler

Taking office in 2013, President Rouhani made it known that improving relations with Saudi Arabia would be high on his list. Ayatollah Rafsanjani, a long-time advocate of better ties with Riyadh, was convinced that with the pending nuclear agreement, the Joint Comprehensive Plan of Action (JCPOA), an overture to the kingdom was in order. However, Rouhani's approach to improving relations with the Arab world has been criticised by conservative rivals determined to undermine him. More detrimental was the growing involvement of IRGC in the civil war in Yemen.

In 2014, the Houthi rebels occupied Sana and expelled the legitimately elected and internationally recognized President Abdrabbuh Mansur Hadi. They were joined by militants loyal to the former president Saleh. After capturing Sana, the Houthis made rapid progress and seized additional

[10] Jiraroy Mamdkul, "Saudi-Arabia-Iran's Foreign Policy Crisis: A Case Study of Execution of Saudi Shiite Cleric Sheikh Nirm al-Nimr," *RJSH* vol. 4, January–June, 2017, 75–82; Gfoeller, Michael (2008-08-23). "Meeting with controversial Shi'a sheikh Nimr". *WikiLeaks*. WikiLeaks cable: 08RIYADH1283. Archived from the original on 2012-01-23, available at https://wikileaks.org/cable/2008/08/08RIYADH1283.html; The Shia in Saudi Arabia The sword unsheathed, *The Economist*, October 18, 2014, https://www.economist.com/news/middle-east-and-africa/21625868-protests-break-out-after-shia-cleric-sentenced-death-sword-unsheathed; TNA, Saudi Arabia Would Pay Dearly if it Carried out the Execution, *Tasnim News Agency*, https://www.tasnimnews.com/fa/news/1393/04/05/413853/; FNA, "Offensive Undergrounds": The IRGC Newest Weapon in the Asymmetric Marine Battles, *Fars News Agency*, April 26, 2011, http://www.farsnews.com/newstext.php?nn=9002031362.

strategic outposts, including the port of al-Hudaydah near Bab al-Mandeb in the Gulf of Aden. A major route for shipping supplies to the rebels, al-Hudaydah was also crucial to a long-term IRGC naval security project. On land, the Houthi rebels threatened cross-border raids into the Eastern Province during their southern offensive. In response, Saudi Arabia began a military buildup on its border. The Houthis responded with defiance, boasting that their troops would counterattack against any Saudi aggression and would not stop until Riyadh was taken.[11]

While it was not clear whether the IRGC ordered or approved the raids, King Salman bin Saud, crowned on January 23, 2015, and his young son Crown Prince Mohammed bin Salman (MBS), the Minister of Defense, blamed Iran for the situation. The energetic Prince Mohammed, who spearheaded the ambitious Vision 2030 project to turn Saudi Arabia into a global powerhouse, viewed Iran as a dangerous enemy on a quest for regional hegemony and an unrepentant destabilizer of the GCC states. Alarmed by the proclamations from Riyadh, the moderates tried to mend the relations. In a rare case of resolve, the government prevailed on the Guards to turn around a ship loaded with supplies for the Houthis. Yalasarat, the IRGC media outlet, wrote that "we have sent a ship of rescue" with 2500 tons of first aid to Yemen, but from the government's top level, it was announced that sending this ship to the port of al-Hodeidah was an unbridled act that should not have been done, and Iran's "ship of rescue" finally docked at Djibouti instead of the Yemeni port of al-Hodeidah.[12]

Rafsanjani announced plans to travel to Riyadh on March 31, 2015, but his trip was derailed when King Salman launched an intervention codenamed Operation Decisive Storm (Amaliyyat `Āṣifat al-Ḥazm) on March 25 against Houthi fighters in Yemen. Supporting the Saudis was a coalition of allied forces—UAE, Bahrain, Kuwait, Jordan, Morocco, Egypt, Qatar, Senegal, and Sudan—which pledged to restore Hadi to his rightful position.[13]

The Saudis and Hadi loyalists accused Iran of helping the Houthis, a charge which American intelligence and other sources supported. A study by Conflict Armament Research (CAR), a London-based organization

[11] Almasmari, Hakim (24 March 2015). "Yemen's Houthi Militants Extend Push Southward". *The Wall Street Journal*. Retrieved 25 March 2015.

[12] Yalasarat, Who in Iran prevented the presence of Iranians in the Hadida port of Yemen? *Yalasarat*, May 25, 2015, yalasarat.com/prtbzzb85rhbgzp.uiur.html.

[13] Situation in Yemen two years after launching Operation Decisive Storm, *Middle East Observer*, March 29, 2017, https://www.middleeastobserver.org/2017/03/29/situation-in-yemen-two-years-after-launching-operation-decisive-storm/.

funded by the European Union (EU) to monitor movement and use of conventional weaponry, indicated the substantial Iranian involvement. IRGC and, uncharacteristically, Artesh provided Houthis with a range of weapons. The Houthis were sent the Qasef-1 UAV (drone) nicknamed "kamikaze" to target coalition missile defense systems in "kamikaze" attacks. The militants used the Ababil drones fitted with high explosive warheads to engage high-value targets, such as radar and Patriot missile batteries.[14]

To help in naval battles, Iran supplied unmanned, remote-controlled boats laden with explosives. The UAE military sources reported that Iran transferred anti-ship and man-portable missiles, including short-range Scud missiles, and surface-to-air missiles reconfigured to operate as surface-to-surface rockets. Also included in the shipments were naval mines. On land, the Houthis had the use of Iranian Dehlavieh anti-tank guided weapons (ATGW) and the Russian-manufactured 9 M133-1 Kornet ATGW.[15]

CAR's analysis of the seized contents and its investigations of the traffic of the traditional boats, the dhows, around the Horn of Africa, suggested the existence of a weapon pipeline extending from Iran to Somalia and Yemen. Local sources in Puntland, Somalia, also report that Iran used the region's ports as transshipment points for weapons. Between February and March 2016, the British warships, HMAS *Darwin*, the French FS *Provence*, and the American USS *Sirocco*, operating as part of the multinational Combined Maritime Forces (CMF), intercepted three dhows transporting weapons in the Arabian Sea. The boats, carrying a significant quantity of military material on board, were thought to have originated in Iran and were destined for Somalia and Yemen.[16]

To avoid interdiction, the Guards began using waters further along the Gulf between Iran and Sur, Oman. Iranian ships transferred the arms to smaller vessels which faced less scrutiny.

The Guards also developed an alternative overland route which took advantage of the porous 288-km long Yemeni-Omani border. The arms smugglers have used Shehen as a point of entry to the disputed al-Mahra

[14] CAR, Frontline Perspective: Iranian Technology Transfers to Yemen, *Conflict Armament Research*, March 2017, http://www.conflictarm.com/download-file/?report_id=2465&file_id=2467.

[15] Ibid.

[16] CAR, Dispatch from the Field, Maritime Interdictions of Weapon Supplies To Somalia And Yemen, Deciphering a link to Iran, Maritime Interdictions of Weapon Supplies to Somalia And Yemen, Deciphering a link to Iran, *Conflict Armament Research*, November 2016.

governorate whose people have been engaged in a struggle for autonomy from the central government in Yemen.[17]

For President Rouhani the well-documented IRGC-QF involvement in the Yemeni Civil War was one more obstacle to normalization. But, like in Syria, the government had little input into the decisions of the parastatals. Qassem Suleimani considered Yemen to be of strategic importance, not least because of Ayatollah Khamenei's support of the 20-year naval plan. As described in Chap. 2, the anti-access, anti-denial (A2/AD) strategy aimed at complicating American efforts to patrol the seas. When the regime was under threat or planning to retaliate, it could wreak havoc in strategic passageways within the triangle stretching from the Straits of Hormuz to Bab al-Mandeb in the Red Sea and the Strait of Malacca in the Indian Ocean. According to a senior Iranian official, Suleimani believed that "Yemen is where the real proxy war is going on and winning the battle in Yemen will help define the balance of power in the Middle East."[18]

Yemen was only one of the tension points that aggravated the relationship between Iran and the kingdom. A stampede during 2015 haj that took the lives of more than 500 Iranian pilgrims, the single largest national contingent among the more than 1,000 victims, added to the acrimony. In a curious twist, Ghazanfar Roknabadi, a former Iranian ambassador to Lebanon who escaped an attack in Beirut in 2012, allegedly by a pro-Saudi militia, was among the dead. The Saudis were late in identifying his body, leading to conspiracy theories in Tehran. More of a threat to Riyadh, however, was Ayatollah Khamenei's take on the disaster. The Supreme Leader accused the Saudi authorities of incompetence and demanded that an Islamic consortium, including Iran, should take over the administration of the haj. Speaking in a televised address, Khamenei warned of repercussion and concluded: "But they should know that Iran's hand is superior to many others and has more capabilities." To the Saudis, who considered themselves the titular guardians of the holy places, the idea of a consortium was a clear sign that the Iranians wanted to usurp their place in the Muslim world, not to mention their desire to achieve a political hegemony in the region.[19]

[17] Jonathan Saul, Exclusive: Iran Revolutionary Guards find new route to arm Yemen rebels, *Reuters*, August 1, 2017, www.reuters.com/article/us-gulf-kuwait-iran-exclusive-idUSKBN1AH4I4.

[18] Reuters, Exclusive: Iran steps up support for Houthis in Yemen's war—sources, *Reuters*, March 21, 2017, https://www.reuters.com/article/us-yemen-iran-houthis/exclusive-iran-steps-up-support-for-houthis-in-yemens-war-sources-idUSKBN16S22R.

[19] Ben Hubard, "The Number of Iranians Killed in Hajj Stampede Grows to 464," *New York Times*, October 1, 2015.

Already angered by the statement of the Supreme Leader, King Salman and his son had to contend with a renewed wave of unrest among the Shiites in the Qatif and Al-Ahssa governorates in the Eastern Province. The intelligence services indicated that Iran was behind the disturbances in which the imprisoned al-Nimr was also implicated. On January 2, 2016, Riyadh announced the execution of Shiite cleric and 46 others who were accused of terror attacks. Saudi and Western sources alleged that the government would not tolerate the destabilization. It was a sign to Iran that, in the struggle for regional hegemony, the Kingdom would not be deterred from taking painful and costly decisions.[20]

Shiites protested in several countries, but the reaction in Iran was quite extreme. On the day of the execution, hundreds of demonstrators in Tehran set fire to the Saudi Embassy before the police made 40 arrests. In Mashhad, north-eastern Iran, four more protesters were arrested after a crowd torched a Saudi consulate. To contain the damage, President Rouhani denounced the attacks on the Saudi Embassy and the consulate, calling them "totally unjustifiable." Rafsanjani also urged for restoring calm, but the normalizers were not successful. Rafsanjani, who died in January 2017, was aware that, when it came to Saudi Arabia, his power is limited. His friend, Ayatollah Vahid Khorasani recalled that Rafsanjani once told him that there was little point in his Saudi mediation project. "What should I say if I go to Saudi Arabia? I have no power to say that I agree with this or disagree with that. This is primarily the responsibility of the Supreme Leader."[21]

And it was the Supreme Leader who was driving much of the animosity with his harsh language. In his words, "God's hand of retaliation will grip the neck of Saudi leaders," and Saudis would face "divine revenge" for the killing of an "oppressed scholar." The Saudis responded by cutting diplomatic relations and asking the Iranian diplomats to leave the Kingdom

[20] FNA, Why did Saudi Arabia Execute Sheikh Nimr?, *Fars News Agency*, January 4, 2016, www.farsnews.com/newstext.php?nn=13941013001649; Alef, Why did Saudi Arabia Execute Sheikh Nimr?, *Alef.ir*, January 3, 2016, old.alef.ir/vdcizzapyt1ary2.cbct.html?321717; Diplomacy Irani, Saudi Calculation in Execution of Sheikh al-Nimr, *Diplomacy Irani*, January 3, 2016, www.irdiplomacy.ir/fa/page/1955272/محاسبات+عربستان+در+اعدام+شیخ+نمر+.html.

[21] ISNA, Memory of a Relationship, *Iranian Students News Agency*, January 10, 2017, www.isna.ir/news/95102113283/خاطرات-یک-رابطه-تصاویر.

within 48 hours. Riyadh also stopped air traffic and trade links between the two countries.[22]

While the attacks seemed spontaneous, they were carefully staged. Hassan Kordmihan, acting on orders from Reza Panahian, a member of the Haqqani School network of Ayatollah Mohammed Taqi Mesbah Yazdi, organized the crowd that stormed the embassy. Meshbah Yazdi was a bitter foe of the Saudis whom he accused of catering to the American and Zionist agenda in the Gulf. Indeed, many protesters in the crowd carried placards denouncing the Big and Small Satan, a reference to the United States and Israel. Soon after, the Tehran municipality proposed naming a street after Sheikh Nimr. In fact, Mohsen Pirhadi, head of Basij at city council of Tehran ordered activists to change the sign on Bostan Street where the Saudi embassy was located to Nimr al-Nimr Street. In a rare public spat, the Ministry of Foreign Affairs forced the Tehran City Council to change the street name back to its original. But the Council scored a victory when it voted to change the name of Naz Street to Sheikh al-Nimr.[23]

Although it is impossible to determine whether the IRGC-QF commanders were involved in planning the assault, Ayatollah Khamenei's continuous vitriol against Riyadh did not bode well for Rouhani and Rafsanjani who were forced to denounce the execution of Sheikh Nimr. More turmoil forced their hand when, on January 7, 2016, the Saudi-led coalition bombed the Iranian embassy in Sana, which wounded several guards.[24]

[22] Townsend M., Sheikh Nimr al-Nimr: Shia cleric was a thorn in Saudi regime's side, *The Guardian*, http://www.theguardian.com/world/2016/jan/02/sheikh-nimr-alnimr-shia-cleric-thorn-saudi-regime-side; Jiraroj Mamadkul, Saudi Arabia – Iran's Foreign Policy Crisis: A Case Study of Execution of Saudi Shia Cleric Sheikh Nimr al-Nimr, *Institute of Diplomacy and International Study, RJSH* Vol. 4, No. 1, January–June 2017, pp. 75–82.

[23] RF, Objection of Ministry of Foreign Affairs to Changing of Bostan Names to Ayatollah Nimr, *Radio Farda*, https://www.radiofarda.com/a/f7-foreign-ministry-objects-over-changing-street-name-to-nimr/27479649.html; Kayhan Newspaper, Changing of Naz Street to Sheikh Nimer, *Kayhan Newspaper*, September 10, 2016, kayhan.ir/fa/news/83580/.

[24] Tabnak, Damage to the Iranian embassy in the Saudi attack on Sanaa, *Tabnak News Agency*, September 20, 2015, www.tabnak.ir/fa/news/532918/آسیب-به-سفارت-ایران-در-حمله-سعودیها-به-صنعا; Entekhab, Iran Warns Saudi Arabia: If you once Again Attack Areas Near the Iranian Embassy in Yemen, You Will Face Serious Consequences, *Entekhab.ir*, www.entekhab.ir/fa/news/209159/هشدار-ایران-به-عربستان-اگر-بار-دیگر-به-مناطق-نزدیک-سفارت-ایران-در-یمن-حمله-کنید-با-پیامدهای-جدی-مواجه-خواهید-شد.

The Iranian Foreign Minister Mohammad Javad Zarif called for "an immediate halt to Saudi Arabia's military operations in Yemen." He added that the attack "would escalate tensions in the region." Rouhani said, "This is not a war, but a cowardice of a neighboring country. This government is a naïve government which does not understand regional and global politics." Ali Shamakhi, the Secretary of the Supreme National Security Council (SNSC), denounced Riyadh for killings civilians and causing a massive humanitarian crisis in Yemen. An Iranian envoy to the United Nations sent a letter denouncing the attack. He added that "I would like to warn that a repetition of similar airstrikes close to my country's diplomatic representation in the future can have serious consequences, including for the safety and security of Iranian diplomats in Sana."[25]

Much as the government tried to follow official diplomatic protest, Ayatollah Khamenei used the occasion to undermine Saudi spiritual hegemony in the Muslim world. In his haj message on September 5, 2016, the Supreme Leader issued a call to boycott the pilgrimage. He urged all Muslim countries to recognize the true nature of the Saudi rulers, who are "criminals, and servants of the American Satan and the Zionists."[26]

To observers acquainted with the vituperative language of the Supreme Leader, there was nothing out the ordinary in his missive. But, as described in the previous chapters, an increase in references to the "American Satan," a phrase that had gone out of use, indicated the Principalists' anxiety about the prospect of a Donald Trump presidency.

[25] Arash Karami, Iran condemns Saudi bombing of Yemen, *Almonitor*, March 26, 2015, www.al-monitor.com/pulse/originals/2015/03/iran-condemn-saudi-attack-bombing-yemen.html; Entekhab, Rouhani: War in Yemen, is a Cowardice attack on a Neighbor, *Entekhab.ir*, May 9, 2015, www.entekhab.ir/fa/news/203079/روحانی-جنگ-یمن-حمله-ناجوانمردانه-همسایه-یک-است-آل-سعود-تازه-کار-و-است-مسائل-سیاسی-را-نمی-فهمد-می-خواهند-فقط-خود-را-نشان-دهند; Afshon Ostovar, Vanguard of the Imam Religion, Politics, and Iran's Revolutionary Guards (London: Oxford University Press) 2015, 4380; IFP, Iran warns Saudi Arabia against targeting areas near Iranian embassy, *Iran Front Page*, June 10, 2015, ifpnews.com/news/world/middle-east/iran-warns-saudi-arabia-against-targeting-areas-near-iranian-embassy/.

[26] JAM News, "Hashemi: 'It Doesn't Hurt if We First Offer Our Hand to Saudi Arabia,'" *JAM News*, September 4, 2016, www.jamnews.ir/detail/News/712527; FNA, "There Is a Blood Sea between Iran and Saudi Arabia," *Fars News Agency*, September 9, 2016, www.farsnews.com/newstext.php?nn=13950619000247; and "Leader of the Revolution; Saudi Arabia Has No Credibility to Administer the Holy Sites," *Fars News Agency*, September 7, 2016, www.farsnews.com/newstext.php?nn=13950617000839.

The Trump Administration: Tilting the Balance Toward Saudi Arabia

For the Saudi government, a change in American policy could not have come soon enough. Bahrain's chronic upheaval worried Riyadh greatly. By 2016, the number of Shiite groups had increased, and several became radicalized. Bahraini authorities listed The February 14 Movement, Saraya al-Ashtar (Ashtar Brigade), Saraya al-Muqawama al-Shabiya (Resistance Brigade, or SMS), Saraya al-Mukhtar (Bahraini Islamic Resistance), Quroob al-Basta and Saraya al-Karar as a security threat. In March 2014, Saraya al-Ashtar and SMS took responsibility for the killing of two Bahraini police officers and an officer from the United Arab Emirates. Others were implicated in planting grenades in a mall and an attack on a power station. While some groups called to "purge" Bahrain of the "Saudi and Khalifa occupiers," others, like Saraya al-Mukhtar wanted to reunite the Bahraini Shiites with their brethren in the Eastern Province.[27]

Both Bahrain and Saudi Arabia blamed Iran for supporting the radical Shiite groups, but the Obama administration was reluctant to accept the charges. To push their case, in the fall of 2016, the Bahraini authorities prepared a dossier documenting Iran's involvement in arming and training the terror groups. The dossier included technical evidence on explosives and bombs forensically linked to Iran as well as a report on the training in guerrilla warfare, which the IRGC provided to Bahraini Shiites including the al-Ashtar Brigades. The Washington Post, which obtained a copy of the dossier, quoted a Western intelligence official admitting to "seeing a new boldness by Iran in supporting armed insurgents in the kingdom." On March 17, 2017, Trump's State Department designated Ahmad Hasan Yusuf and Alsayed Murtadha Majeed Ramadhan Alawi, the two leaders of the al-Ashtar Brigades, as "global terrorists" and specifically accused Iran of backing the group as part of its "destabilizing and terrorism-related activities in the region."[28]

[27] Giorgio Cafiero, "Bahrain's Ongoing Political Crisis Threatens Stability," *Al-Monitor*, May 21, 2017.

[28] Souad Mekhennet and Joby Warrick, "U.S. increasingly sees Iran's hand in the arming of Bahraini militants," *Washington Post*, April 1, 2017, https://www.washingtonpost.com/world/national-security/us-increasingly-sees-irans-hand-in-the-arming-of-bahraini-militants/2017/04/01/be5e61fc-1329-11e7-833c-503e1f6394c9_story.html?utm_term=.47f77529d33a; Kylie Atwood, State Department adds two men to

Having scored a major victory in linking Iran to the unrest in Bahrain, the Saudi authorities were eager to push the narrative further. In several interviews and speeches in Washington, Mohammed bin Salman accused Iran of systematically destabilizing and radicalizing the region. In a departure from traditional Saudi policy, he threatened that the future battlefield would be inside Iran and not Saudi Arabia. The blunt pronouncement drew a sharp response from Brigadier General Hossein Dehghan, the Iranian Minister of Defense, who questioned Saudi capability: "They think they can do something because of having an air force." Amir Ali Hajizadeh, the air force commander of Guards, remarked that "should the enemy make a mistake, our roaring missiles will rain down on them."[29]

But unlike under Obama, where such outbursts frequently went unanswered, the Trump administration came down hard on the IRGC. In his March 2017 testimony to Congress, General Joseph L. Votel, the head of CENTCOM, listed Iran as one of the "drivers" of instability and referred to its "malign influence" in Syria, Iraq, and other hot spots. On a different occasion, the CENTCOM chief declared that "Iran posed the greatest threat to long-term stability" in the region and was seeking hegemony through its proxies. Votel expressed concern that after the fall of ISIS, Iran would be poised to expand its reach. Votel, who was on record denouncing the behavior of the IRGC naval unit (the Niruye Daryai Sepah-e Iran or NEDSA) in the Straits of Hurmuz, as noted in Chap. 2, made it clear that the American navy would be deployed to help Saudi Arabia in the Gulf of Aden.[30]

The Iranian high command responded aggressively. Dehghan urged the U.S. Navy to leave the Gulf region and Rear Admiral Ali Fadavi, the commander of the IRGC naval division, announced that the "martyrdom-seeking combatants" in speedboats would be deployed against American navy. Fadavi, who previously boasted that the Iranians can destroy American ships, added that the martyrdom operations were a "miracle" that would deter the Americans. Rear Admiral Habibollah Sayyari, the

global list of terrorists, *CBS News*, March 17, 2017, https://www.cbsnews.com/news/state-department-adds-two-men-to-global-list-of-terrorists/.

[29] TNA, Dehghan promised that Iran would destroy Saudi Arabia except for Mecca and Medina. *Tasnim News Agency*, May 8, 2017; IRNA, The Guards' Air Force Commander: The New Generation of Long-range Missiles are on their Way to Come, *IRNA*, www.irna.ir/fa/News/81801053/.

[30] Joseph L. Votel, "The Posture of U.S. Central Command, *Senate Armed Service Committee*, March 9, 2017, https://www.armed-services.senate.gov/imo/media/doc/Votel_03-09-17.pdf.

commander of the regular Iranian Navy (Islamic Republic of Iran Navy or IRIN), disclosed that the 45th Group—a ship and a destroyer—left for the Gulf of Aden and Bab al-Mandeb in March to augment the IRGC naval units. Sayyari explained that he acted according to the long-term plan to increase Iranian presence in international waters.[31]

If the Iranian hardliners nurtured hopes of deterring Washington with a show of naval prowess, they were bound to be disappointed. On May 20, 2017, President Trump arrived in Riyadh on a visit, which, by most estimates, was an effort to bolster Saudi Arabia and its Sunni partners and isolate Iran, in addition to fighting Islamic extremism. Saudi Arabia scored another victory when Trump added Iran to the list of radical Islamists and promised a partnership to fight it. Officials involved in the trip explained that there were two forms of extremism: the self-styled jihadists, such as ISIS; and the Iranian-led consortium of subnational groups, militias, and other proxies. When addressing the conference of 55 Islamic countries, which the Saudis hosted, the president implied that Iran should be treated like ISIS. He delivered the same message to a meeting of the GCC and, according to sources, approved the Saudis' plans to take on Iran during a secret bilateral meeting. A White House statement noted that Trump and King Salman had directed their teams to find ways "to further strengthen and elevate the United States–Saudi strategic relationship" on the political, military, security, economic, cultural, and social fronts. The two sides agreed to increase cooperation in the energy field and start a new U.S.-Saudi program "worth potentially more than $200 billion in direct and indirect investments within the next four years." The two sides agreed on another $350 billion investment over the next ten years.[32]

[31] TNA, Iran's Naval Patroling at the Entrance of the Bab al-Mandeb Strait, *Tasnim News Agency*, April 30, 2015, https://www.tasnimnews.com/fa/news/1394/02/10/726907/المندب-باب-تنگه-ورودی-در-ارتش-دریایی-نیروی-ناوهای-زنی-گشت; FNA, Iran Navy will be Available Wherever Required, *Fars News Agency*, April 8, 2015, www.farsnews.com/newstext.php?nn=13940119000077; Ahmad Majidyar, Iran Dispatches Flotilla of Warships to Gulf of Aden, *The Middle East Institute*, April 17, 2017, www.mei.edu/content/io/iran-dispatches-flotilla-warships-gulf-aden.

[32] Readout of the President's Meeting with Mohammed bin Salman Abdulaziz Al Saud, Deputy Crown Prince and Minister of Defense of the Kingdom of Saudi Arabia, *The White House Office of the Press Secretary*, March 15, 2017, https://www.whitehouse.gov/the-press-office/2017/03/15/readout-presidents-meeting-mohammed-bin-salman-abdulaziz-al-saud-deputy.

During the trip, the president signed a multi-billion military aid package. The package included the Lockheed Martin-produced Terminal High Altitude Arial Defense (THAAD) missile defense system worth $1 bn. Also included was the Command and Control, Battle Management, Communications (C2BMC) software system, which the American Missile Defense Agency (MDA) used to integrate ballistic missile defenses. Also included were littoral combat ships, armored personnel carriers, missiles, bombs, and munitions. Four multi-mission surface combatant ships worth $11.5 billion, which the State Department approved for purchase in 2015, were added to the contract. The administration sold $1 billion worth of munitions, including armor-piercing Penetrator warheads and Pave way laser-guided bombs. The items were part of an order which President Obama had suspended out of concern for Tehran. The transaction also included an unspecified number of Bradley vehicles and M109 Paladin 155 m howitzers.[33]

The *Washington Post* reported that some of the production and assembly would be in Saudi Arabia, helping to build an indigenous defense industry. For a close watcher of the region, the unprecedented event signaled the beginning of a process to create the so-called "Arab NATO." Long bandied by some American military leaders such as John Allen, and John Stavridis, the former NATO commander, the proposed organization would feature Sunni countries organized as a "response force" of the Arab League. Reports suggest that the initial force would be composed of some 40,000 personnel from a variety of Arab nations—mostly from Egypt, Jordan, Morocco, Saudi Arabia, and Sudan. The United States was expected to play a support role. While the Arab NATO was to be based in Egypt, it would be commanded by a Saudi general. Like NATO, the new alliance will have specified warfighting components: up to 1000 servicemen in the air command; 5000 service members would be assigned to the naval command, and roughly 35,000 will be part of the land forces.[34]

[33] Javier E. David, US-Saudi Arabia seal weapons deal worth nearly $110 billion immediately, $350 billion over ten years, *CNBC*, May 20, 2017, www.cnbc.com/2017/05/20/us-saudi-arabia-seal-weapons-deal-worth-nearly-110-billion-as-trump-begins-visit.html; Ali Vitali, U.S.-Saudi Arabia Sign More Than $110B Arms Deal Amid Trump Visit, *NBC News*, May 20, 2017, www.nbcnews.com/storyline/trumps-first-foreign-trip/u-s-saudi-arabia-sign-110b-arms-deal-plus-another-n762546.

[34] Josh Rogin, Trump to unveil plans for an 'Arab NATO' in Saudi Arabia, *The Washington Post*, May 17, 2017, https://www.washingtonpost.com/news/josh-rogin/wp/2017/05/17/trump-to-unveil-plans-for-an-arab-nato-in-saudi-arabia/?utm_

Few details emerged from the closed-door meetings between American and Khalid bin Ali al-Humaidan, the chief of the General Intelligence Directorate (GID). Several issues, however, were apparently on the agenda, according to Western intelligence sources. One item pertained to a new evaluation of the war in Yemen. As already noted, the Obama administration was reluctant to support the Saudi-led intervention, not least because of its policies of protecting its JCPOA agreement. Human rights groups which carried weight in Washington decried the civilian toll of Saudi air strikes. The Americans have used drone attacks against al-Qaeda in the Arabian Peninsula (AQAP), but struck against the Houthis and the Iranians only once in retaliation to the failed missile attacks on the U.S. Navy ship USS *Mason* by Houthis on October 12, 2016. The Pentagon fired several Tomahawk cruise missiles and destroyed three coastal radar sites in areas of Yemen controlled by Houthi forces.[35]

Restoring the legitimate government in Yemen has been a high priority for the Trump administration, which was aware of the Iranian anti-access/area-denial (A2/AD) plan. As head of CENTCOM, Gen. James Mattis warned in 2011 about the Iranian plans to disrupt American naval presence in the region. While American air and naval bases—Masira Base, Diego Garcia Base, Canso Base, and Massawa Base, among them—were seemingly not susceptible to A2/AD, the Pentagon took the threat seriously.[36]

Another issue on the bilateral U.S.-Saudi agenda included a more robust effort to increase Iran's "cost of doing business," a rational choice term applied counterterrorism. Briefly, the notion called for radicalizing ethnic and sectarian sentiments in Iran. According to this theory, should Iran become preoccupied with its internal problems, it would have fewer resources to deploy in its quest for hegemony. As if on cue, on June 2, 2017, the Central Intelligence Agency announced the creation of a new

term=.0e3f4b8b0681; James Stavridis, "The Arab NATO," *Foreign Policy*, April 9, 2015, foreignpolicy.com/2015/04/09/the-arab-nato-saudi-arabia-iraq-yemen-iran/.

[35] Phil Stewart, U.S. military strikes Yemen after missile attacks on U.S. Navy ship, *Reuters*, October 12, 2016, https://www.reuters.com/article/us-yemen-security-missiles/u-s-military-strikes-yemen-after-missile-attacks-on-u-s-navy-ship-idUSKCN12C294.

[36] Stephen Hughes, Yemen, Iran's strategic naval expansion, anti-access/area denial (A2/AD) Part I of III, *Jerusalem Post*, April 28, 2015, www.jpost.com/Blogs/The-Iran-Threat/Yemen-Irans-Strategic-Naval-Expansion-anti-accessarea-denial-A2AD-Part-I-of-III-399400.

Iran Center under Michael D'Andrea, the former head of the Agency's Counterterrorism Center.[37]

One possible option was to support militant groups in Pakistani Baluchistan operating across the Iranian borders in the province of Sistan and Baluchistan. Still, experts have warned that launching terror attacks inside Iran would be an uphill battle because The Ministry of Intelligence and Security (MOIS) and the IRGC have been diligent in fighting domestic terrorism. In May, Heydar Moslehi, the MOIS chief, announced that his office had dismantled 30 terror cells, including some Kurdish ones. Moslehi stated that the Kurdish groups received support from Saudi Arabia. The June 7 terror attack on the parliament and the mausoleum of Ayatollah Khomeini represented the most serious setback to the security forces since the bombings of the 1980s. In fact, observers credited this sophisticated apparatus for shielding Iran from the type of high-profile attacks ISIS has inflicted in European countries.[38]

Hardliners were quick to blame the Rouhani government. They mocked the president for saying that he had removed the shadow of war from Iran through the nuclear deal. The *Kayhan* newspaper, reflecting the thinking of the Supreme Leader, attributed the assault on Rouhani's efforts to undermine the IRGC. The daily wrote: "When you [Rouhani] slammed and weakened the Revolutionary Guards for testing missiles when you questioned all of the past 38 years, it gave a good opportunity to America, Israel, and Saudi Arabia to fill the gap with security challenges." Without producing evidence, hardliners were quick to blame Saudi Arabia, a country, which, in their opinion, was involved, along with Israel, in training the terrorists who assassinated several nuclear scientists in the early 2010s.[39]

[37] Matthew Rosenberg and Adam Goldman, C.I.A. Names the 'Dark Prince' to Run Iran Operations, Signaling a Tougher Stance, June 2, 2017, https://www.nytimes.com/2017/06/02/world/middleeast/cia-iran-dark-prince-michael-dandrea.html.

[38] Farhad Rezaei, Trump's Regime Change Strategy toward Tehran is Doomed to Fail, *Center for Iranian Studies*, June 10, 2017, iramcenter.org/en/trumps-regime-change-strategy-toward-tehran-is-doomed-to-fail/; Farhad Rezaei, Terrorist Attacks in Iran Could Strengthen Hardliners, *Atlantic Council*, June 9, 2017, www.atlanticcouncil.org/blogs/iraninsight/terrorist-attacks-in-iran-could-strengthen-hardliners.

[39] Hamshahri, The List of Assassinations of Iranian Nuclear Scientists Over the Past two Years; Mossad Carried out Terrors, *Hamshahri Online*, www.hamshahrionline.ir/details/156984; FNA, Failure Terror Triangle, *Fars News Agency*, www.farsnews.com/newstext.php?nn=13910516000392; Afshon Ostovar, Vanguard of the Imam Religion, Politics, and Iran's Revolutionary Guards (London: Oxford University Press, 2015), 4380.

Iranian involvement in the Syrian civil war was another issue which the Saudis and the Americans had apparently reviewed. At the onset of the war, the Saudis, sworn enemies of Assad, spent considerable resources to defeat him and his Iranian protectors. Prince Bandar bin Sultan, the then head of the GID, devised a plan to use proxy groups against Iranians in and outside of Iran. The Iranian intelligence services had blamed Prince Bandar for the 2013 attack on the Iranian embassy in Beirut, which targeted the ambassador Ghazanfar Roknabadi. The ambassador escaped the attack, which killed at least 23 people, including cultural attaché Sheikh Ebrahim, and wounded more than 150. Roknabadi escaped but, as noted above, died in the haj stampede. The Sunni Abdullah Azzam Brigades claimed responsibility for the attack to avenge the Iranian-sponsored militias in Syria.[40]

Prince Bandar was dismissed on April 15, 2014, and the Saudi involvement in Syria gradually declined. But Prince Mohammed, who sought to revive Riyadh's influence, sent the intelligence chief al-Humaidan to Jordan and Israel. While allegations about intelligence contacts between Israel and Saudi Arabia are not new, the more recent meetings involved coordinating efforts to push back against the IRGC-QF. The Saudis were concerned about a possible Iranian infringement on the Jordanian border. The intelligence chief also discussed the so-called "Riyadh First" project based in the al-Ghadi base under the IRGC Air Force Commander General Amir Ali Hajizadeh.[41]

The Iranians allegedly added an extra 100 km to the intermediate range Scud-C (500 km) and Scud-D (700 km) surface missiles to attack Saudia Arabia. Indeed, on February 4, 2017, the Houthis fired the modified missile Scud-D, which hit the al-Muzahimiyah military camp west of Riyadh. Soon after, the Houthis boasted that they were in possession of missiles that could reach Israel. The Houthis accused Israel of providing Saudi Arabia with intelligence collected by its reconnaissance ships based in the port of Assab, Eritrea.[42]

[40] FNA, Who is Behind Israel Secret War Against Iran, December 8, 2013, *Fars News Agency*, www.farsnews.com/13920913001234; Iran Times, Bombers attack Iran embassy in Beirut, *Iran Times*, November 22, 2013, iran-times.com/bombers-attack-iran-embassy-in-beirut/.

[41] Sputnik, Israeli Media: Saudi Spy Chief Secretly Visited Israel, Palestinian Autonomy, *Sputnik*, February 25, 2017, https://sputniknews.com/politics/201702251051027892-saudi-spy-chief-secret-meetings/.

[42] PressTV, Ballistic missile attack reported in Saudi capital: Yemeni media, *PressTV*, February 6, 2017, presstv.ir/DetailFr/2017/02/06/509301/yemen-saudi-arabia-riyadh-

The Trump visit produced some unintended consequences as well, at least from the American perspective. Feeling empowered, the Saudis decided to challenge Qatar, whose policy of supporting the Muslim Brotherhood and closeness with Iran had rankled Riyadh for years. In an unprecedented move, in June 2017, Saudi Arabia and its allies, including UAE, Bahrain, and Egypt, accused Qatar of supporting terrorism, and acting as a proxy for Iran to destabilize the region. After Qatar refused to submit to a list of stipulations, the coalition broke off relations with Doha and imposed a travel and trade ban. The Americans, who have a large air base in Qatar, tried to mediate but failed to bridge the gap, leading to a new alignment in the region.

The Saudi move against Qatar was a victory for Iran, which shares the world's largest reservoir of natural gas with Doha. Iran quickly helped Qatar overcome the sanctions by shipping and airlifting supplies to Doha. On August 9, 2017, Qatar's Ministry of Interior announced that his country would allow visa-free entry for Iranian citizens. Sheikh Tamim bin Hamad al-Thani, Qatar's emir, called on GCC countries to negotiate with Iran. Qatari officials made it clear that isolating Tehran was a foolhardy policy, which would only help the Trump administration. Moreover, Doha joined Tehran and Ankara in a tripartite agreement to broaden economic and security cooperation. The Doha agreement could not have come at a better time for Iran, which saw to strengthen relations with Ankara.

Conclusion

Saudi Arabia has had a long history of strife with Iran. Immediately after seizing power in 1979, the new regime, intent on exporting its revolution and undermining the Sunni kingdom, launched operations against Riyadh and its Gulf neighbors. In its latest venture, Iran has promoted the Houthi rebellion in Yemen, arguably the most direct challenge to Saudi interest in decades. Riyadh has also taken a dim view of Iran's steadfast support for Bashar al-Assad in Syria, and its meddling in Iraq and Bahrain. Unlike his predecessor, King Salman vowed to respond strongly to Iranian provocations, a decision which led to the Saudi intervention in the civil war in Yemen.

ballistic-missile; Sputnik, Israeli Media: Saudi Spy Chief Secretly Visited Israel, Palestinian Autonomy, *Sputnik*, February 25, 2017, https://sputniknews.com/politics/201702251051027892-saudi-spy-chief-secret-meetings/.

The government of Hassan Rouhani, backed by Ayatollah Akbar Hashemi Rafsanjani, hoped to improve the relationship between the two regional rivals. Rafsanjani was regarded as the only politician capable of nudging Iran towards a détente, a policy which called for a withdrawal from Yemen and eschewing any meddling in in Gulf States. The hardliners utilized the peculiarities of the negotiated political order to undermine the Rouhani government.

Following the death of Rafsanjani, the Rouhani government has struggled to maintain the outreach to Saudi Arabia and striking a note of conciliation, invited Riyadh to collaborate to end the conflicts in Syria and Yemen. However, there are strong indications that the hardliners in the Revolutionary Guards, egged on by the Haqqani clergy, are determined to press on.

References

Afshon Ostovar, Vanguard of the Imam Religion, Politics, and Iran's Revolutionary Guards (London: Oxford University Press, 2015).

Ahmad Majidyar, Iran Dispatches Flotilla of Warships to Gulf of Aden, *The Middle East Institute*, April 17, 2017, www.mei.edu/content/io/iran-dispatches-flotilla-warships-gulf-aden.

Alef, Why Saudi Arabia Executed Sheikh Nmer? *Alef.ir*, January 3, 2016, old.alef.ir/vdcizzapyt1ary2.cbct.html?321717.

Ali Vitali, U.S.-Saudi Arabia Sign More Than $110 B Arms Deal Amid Trump Visit, *NBC News*, May 20, 2017, www.nbcnews.com/storyline/trumps-first-foreign-trip/u-s-saudi-arabia-sign-110b-arms-deal-plus-another-n762546.

Almasmari, Hakim (24 March 2015). "Yemen's Houthi Militants Extend Push Southward," *The Wall Street Journal*. Retrieved 25 March 2015.

Arash Karami, Iran condemns Saudi bombing of Yemen, *Almonitor*, March 26, 2015, www.al-monitor.com/pulse/originals/2015/03/iran-condemn-saudi-attack-bombing-yemen.html.

BBC, Saudi Shia cleric Nimr al-Nimr 'sentenced to death', *BBC News*, October 15, 2014, www.bbc.com/news/world-middle-east-29627766.

Ben Hubard, "The Number of Iranians Killed in Hajj Stampede Grows to 464," *New York Times*, October 1, 2015.

CAR, Dispatch from the Field, Maritime Interdictions of Weapon Supplies To Somalia And Yemen, Deciphering a link to Iran, Maritime Interdictions of Weapon Supplies to Somalia And Yemen, Deciphering a link to Iran, *Conflict Armament Research*, November 2016.

CAR, Frontline Perspective: Iranian Technology Transfers to Yemen, *Conflict Armament Research*, March 2017, http://www.conflictarm.com/download-file/?report_id=2465&file_id=2467.

Chang-Cheng Liu, Saudi-Iranian Relations, 1977–1997, Durham University.

Diplomacy Irani, Saudi Calculation in Execution of Sheikh al-Nimr, *Diplomacy Irani*, January 3, 2016, http://www.irdiplomacy.ir/fa/page/1955272/محاسبات عربستان+در+اعدام+شیخ+نمر++.html.

Entekhab, Iran Warns Saudi Arabia: If you once Again Attack Areas Near the Iranian Embassy in Yemen, You Will Face Serious Consequences, *Entekhab.ir*, www.entekhab.ir/fa/news/209159/هشدار-ایران-به-عربستان-اگر-بار-دیگر-به-مناطق-نزدیک-به- سفارت-ایران-در-یمن-حمله-با-کنید-پیامدهای-جدی-مواجه-خواهید-شد.

Entekhab, Rouhani: War in Yemen, is a Cowardice attack on a Neighbor, *Entekhab.ir*, May 9, 2015, www.entekhab.ir/fa/news/203079/روحانی-جنگ-یمن-حمله-ناجوانمردانه- یک-همسایه-است-آل-سعود-تازه-کار-و-است-مسائل-سیاسی-را-نمی-فهمد-می-خواهند-فقط-خود-را-نشان-دهند.

Farhad Rezaei, Terrorist Attacks in Iran Could Strengthen Hardliners, *Atlantic Council*, June 9, 2017, www.atlanticcouncil.org/blogs/iraninsight/terrorist-attacks-in-iran-could-strengthen-hardliners.

Farhad Rezaei, Trump's Regime Change Strategy toward Tehran is Doomed to Fail, *Center for Iranian Studies*, June 10, 2017, iramcenter.org/en/trumps-regime-change-strategy-toward-tehran-is-doomed-to-fail/.

FNA, "Leader of the Revolution: Saudi Arabia Has No Credibility to Administer the Holy Sites," *Fars News Agency*, September 7, 2016, www.farsnews.com/newstext.php?nn=13950617000839.

FNA, "Offensive Undergrounds": The IRGC Newest Weapon in the Asymmetric Marine Battles, *Fars News Agency*, April 26, 2011, http://www.farsnews.com/newstext.php?nn=9002031362.

FNA, "There Is a Blood Sea between Iran and Saudi Arabia," *Fars News Agency*, September 9, 2016, www.farsnews.com/newstext.php?nn=13950619000247.

FNA, Failure Terror Triangle, *Fars News Agency*, www.farsnews.com/newstext.php?nn=13910516000392.

FNA, Iran Navy will be Available Wherever Required, *Fars News Agency*, April 8, 2015, www.farsnews.com/newstext.php?nn=13940119000077.

FNA, Who is Behind Israel Secret War Against Iran, December 8, 2013, *Fars News Agency*, www.farsnews.com/13920913001234.

FNA, Why Saudi Arabia Executed Sheikh Nimr? *Fars News Agency*, January 4, 2016, www.farsnews.com/newstext.php?nn=13941013001649.

Gfoeller, Michael (2008-08-23). "Meeting with controversial Shi'a sheikh Nimr". *WikiLeaks*. WikiLeaks cable: 08RIYADH1283. Archived from the original on 2012-01-23, available at https://wikileaks.org/cable/2008/08/08RIYADH1283.html.

Giorgio Cafiero, "Bahrain's Ongoing Political Crisis Threatens Stability," *Al-Monitor*, May 21, 2017.

Hamshahri, The List of Assassinations of Iranian Nuclear Scientists Over the Past two Years Mossad Carried out Terrors, *Hamshahri Online*, www.hamshahrionline.ir/details/156984.

IFP, Iran warns Saudi Arabia against targeting areas near Iranian embassy, *Iran Front Page*, June 10, 2015, ifpnews.com/news/world/middle-east/iran-warns-saudi-arabia-against-targeting-areas-near-iranian-embassy/.

Irantimes, Bombers attack Iran embassy in Beirut, *Iran Times*, November 22, 2013, iran-times.com/bombers-attack-iran-embassy-in-beirut/.

IRNA, The Guards' Air Force Commander: The New Generation of Long-range Missiles are on their Way to Come, *IRNA*, www.irna.ir/fa/News/81801053/.

ISNA, Memory of a Relationship, *Iranian Students News Agency*, January 10, 2017, www.isna.ir/news/95102113283/خاطرات-یک-رابطه-تصاویر.

JAM News, "Hashemi: 'It Doesn't Hurt if We First Offer Our Hand to Saudi Arabia,'" *JAM News*, September 4, 2016, www.jamnews.ir/detail/News/712527.

James Stavridis, "The Arab NATO," *Foreign Policy*, April 9, 2015, foreignpolicy.com/2015/04/09/the-arab-nato-saudi-arabia-iraq-yemen-iran/.

Javier E. David, US-Saudi Arabia seal weapons deal worth nearly $110 billion immediately, $350 billion over ten years, *CNBC*, May 20, 2017, www.cnbc.com/2017/05/20/us-saudi-arabia-seal-weapons-deal-worth-nearly-110-billion-as-trump-begins-visit.html.

Jiraroj Mamadkul, Saudi Arabia – Iran's Foreign Policy Crisis: A Case Study of Execution of Saudi Shia Cleric Shaikh Nimr al-Nimr, *Institute of Diplomacy and International Study, RJSH* Vol. 4, No. 1, January–June 2017.

Jiraroy Mamdkul, "Saudi-Arabia-Iran's Foreign Policy Crisis: A Case Study of Execution of Saudi Shiite Cleric Shaikh Nirm al-Nimr," *RJSH* vol. 4, January–June, 2017, 75–82.

Jonathan Saul, Exclusive: Iran Revolutionary Guards find new route to arm Yemen rebels, *Reuters*, August 1, 2017, www.reuters.com/article/us-gulf-kuwait-iran-exclusive-idUSKBN1AH4I4.

Joseph L. Votel, The Posture of U.S. Central Command, *Senate Armed Service Committee*, March 9, 2017, https://www.armed-services.senate.gov/imo/media/doc/Votel_03-09-17.pdf.

Josh Rogin, Trump to unveil plans for an 'Arab NATO' in Saudi Arabia, *The Washington Post*, May 17, 2017, https://www.washingtonpost.com/news/josh-rogin/wp/2017/05/17/trump-to-unveil-plans-for-an-arab-nato-in-saudi-arabia/?utm_term=.0e3f4b8b0681.

Kayhan Newspaper, Changing of Naz Street to Sheikh Nimer, *Kayhan Newspaper*, September 10, 2016, kayhan.ir/fa/news/83580/.

Kylie Atwood, State Department adds two men to global list of terrorists, *CBS News*, March 17, 2017, https://www.cbsnews.com/news/state-department-adds-two-men-to-global-list-of-terrorists/.

Louisa Loveluck, Nimr al-Nimr: Saudi Arabia's outspoken cleric, *Telegraph*, January 2, 2016, www.telegraph.co.uk/news/worldnews/middleeast/saudi-arabia/12078497/Nimr-al-Nimr-Saudi-Arabias-outspoken-cleric.html.

Matthew Rosenberg and Adam Goldman, C.I.A. Names the 'Dark Prince' to Run Iran Operations, Signaling a Tougher Stance, June 2, 2017, https://www.nytimes.com/2017/06/02/world/middleeast/cia-iran-dark-prince-michael-dandrea.html.

Phil Stewart, U.S. military strikes Yemen after missile attacks on U.S. Navy ship, *Reuters*, October 12, 2016, https://www.reuters.com/article/us-yemen-security-missiles/u-s-military-strikes-yemen-after-missile-attacks-on-u-s-navy-ship-idUSKCN12C294.

PressTV, Ballistic missile attack reported in Saudi capital: Yemeni media, *PressTV*, February 6, 2017, presstv.ir/DetailFr/2017/02/06/509301/yemen-saudi-arabia-riyadh-ballistic-missile.

Readout of the President's Meeting with Mohammed bin Salman Abdulaziz Al Saud, Deputy Crown Prince and Minister of Defense of the Kingdom of Saudi Arabia, *The White House Office of the Press Secretary*, March 15, 2017, https://www.whitehouse.gov/the-press-office/2017/03/15/readout-presidents-meeting-mohammed-bin-salman-abdulaziz-al-saud-deputy.

Reuters, Exclusive: Iran steps up support for Houthis in Yemen's war – Sources, *Reuters*, March 21, 2017, https://www.reuters.com/article/us-yemen-iran-houthis/exclusive-iran-steps-up-support-for-houthis-in-yemens-war-sources-idUSKBN16S22R.

RF, Objection of Ministry of Foreign Affairs to Changing of Bostan Names to Ayatollah Nimr, *Radio Farda*, https://www.radiofarda.com/a/f7-foreign-ministry-objects-over-changing-street-name-to-nimr/27479649.html.

Souad Mekhennet and Joby Warrick, "U.S. increasingly sees Iran's hand in the arming of Bahraini militants," *Washington Post*, April 1, 2017, https://www.washingtonpost.com/world/national-security/us-increasingly-sees-irans-hand-in-the-arming-of-bahraini-militants/2017/04/01/be5e61fc-1329-11e7-833c-503e1f6394c9_story.html?utm_term=.47f77529d33a.

Sputnik, Israeli Media: Saudi Spy Chief Secretly Visited Israel, Palestinian Autonomy, *Sputnik*, February 25, 2017, https://sputniknews.com/politics/201702251051027892-saudi-spy-chief-secret-meetings/.

Stephen Hughes, Yemen, Iran's strategic naval expansion, anti-access/area denial (A2/AD) Part I of III, *Jerusalem Post*, April 28, 2015, www.jpost.com/Blogs/The-Iran-Threat/Yemen-Irans-Strategic-Naval-Expansion-anti-accessarea-denial-A2AD-Part-I-of-III-399400.

Tabnak, Damage to the Iranian embassy in the Saudi attack on Sanaa, *Tabnak News Agency*, September 20, 2015, www.tabnak.ir/fa/news/532918/آسیب-به-سفارت-ایران-در-حمله-سعودیها-به-صنعا.

Tali Rachel Grumet, New Middle East Cold War: Saudi Arabia and Iran's Rivalry (California: University of Denver, 2015).

TNA, Dehghan promised that Iran would destroy Saudi Arabia except for Mecca and Medina, *Tasnim News Agency*, May 8, 2017.

TNA, Iran's Naval Patroling at the Entrance of the Bab al-Mandeb Strait, *Tasnim News Agency*, April 30, 2015, https://www.tasnimnews.com/fa/news/1394/02/10/726907/گشت-زنی-ناوهای-نیروی-دریایی-ارتش-در-ورودی-تنگه-باب-المندب.

TNA, Saudi Arabia Would Pay Dearly if it Carried out the Execution, *Tasnim News Agency*, https://www.tasnimnews.com/fa/news/1393/04/05/413853/.

Townsend M., Sheikh Nimr al-Nimr: Shia cleric was a thorn in Saudi regime's side, *The Guardian*, http://www.theguardian.com/world/2016/jan/02/sheikh-nimr-alnimr-shia-cleric-thorn-saudi-regime-side.

Vali Nasr; Tali Rachel Grumet, New Middle East Cold War: Saudi Arabia and Iran's Rivalry (California: University of Denver, 2015).

Youssef M. Ibrahim, Iran Calls U.S. Action A 'Barbaric Massacre', *The New York Times*, July 4, 1988, www.nytimes.com/1988/07/04/world/iran-calls-us-action-a-barbaric-massacre.html.

CHAPTER 8

Iran and Turkey: Frenemies for Ever?

Iran and Turkey, heirs to the two ancient empires, the Persian and the Ottoman, had a history of complicated but non-violent relations. Commentators often noted that they were "bitter friends, bosom rivals." International Relations (IR) theory described this relation as "frenemies," a portmanteau of friend and enemy; that is, an association between two political entities, which form positive tactical bonds despite a fundament strategic dislike. "In the nation-state context, frenemies are able to straddle the gray area between adversity and alliance, and can concurrently castigate and embrace one another." Graphically, the relationship is represented by a sinusoidal cycle that follows a path of somewhat unpredictable ups and downs.[1]

After the Islamic revolution in 1979, the relations between the two countries took a turn for the worse. Ayatollah Ruhollah Khomeini was highly suspicious of Turkey, a secular state, and a NATO member. In a speech on August 24, 1986, Khomeini asserted that "in the Islamic world, the ulama [clergy] were led to believe that they had to obey the tyrants, oppressors, and the holders of naked power. Certain lackeys preferred to obey Atatürk, who destroyed the rule of Islam, instead of obeying the orders of the prophet. How can a reasonable mind accept this? Today, the

[1] Merv Tahiroglou and Behnam Ben Taleblu, "Turkey and Iran. The Best of Frenemies, *Turkish Policy Quarterly*, Vol. 14, No. 3, Spring 2015, 123–134; "Saeid Jafari, "Why Iran–Turkey Tensions Would not Spiral Out of Control," *Al-Monitor*, March 13, 2017; Crisis Group, Briefing No. 51, Middle East and North Africa, December 13, 2016.

F. Rezaei, *Iran's Foreign Policy After the Nuclear Agreement*, Middle East Today, https://doi.org/10.1007/978-3-319-76789-5_8

ulama [in Turkey] who are the puppets of the Pharaonic forces, teach the people the orders of God and the prophet, but at the same time call on them to obey Atatürk…How can one argue that this is consistent with the notion of [Islamic rulers] whom God ordered us to obey? Obviously, [Islamic rulers] in the real sense can only be those who follow the order of God and his messenger."[2]

Khomeini's ideological view affected Iran's diplomatic protocols in Ankara. For instance, during his summer 1987 visit, Prime Minister Mir Hossein Mousavi criticized the Atatürk reforms and refused to pay respect at his mausoleum, a protocol requirement for visiting dignitaries. Mousavi stated that he would prefer to visit the memorial of Mawlana Jalaluddin al-Rumi, an influential Turkish–Iranian theologian in the thirteenth century, located in Konya. Iranian media, which covered the visit, wrote offensive articles about Atatürk. The Turkish media responded in a measure by criticizing Khomeini and his brand of Islamism. In another case, on the 50th anniversary of Atatürk's death in November 1988, the Iran's embassy in Ankara declined to lower its flag to half-mast. The Turkish officials and media described this action as "unforgivable insolence."[3]

To neorealists such as Akbar Hashemi Rafsanjani who was still seared by the devastating war with Iraq, Khomeini's sensibilities were ill-advised. From a practical stand, Ankara turned out to be a good friend in need. The Turks helped Iran to beat the weapons embargo and other sanctions, which the United States imposed to support Saddam Hussein. Iran obtained some of its arms from Turkey and others smuggled through the border. As one newspaper wrote, "during the eight-year war, given the insecurity in the Persian Gulf, we had received many of our requirements from Turkey, or through Turkey's route."[4]

Both moderate presidents, Hashemi Rafsanjani and Mohammed Khatami, supported the practical, mutually beneficial relations with Turkey. But the occasional intrusion of idealism, a subject discussed in Chap. 1, managed to disrupt this equilibrium. Nowhere was this clearer than in

[2] Quoted in Ergun Ozbudun, "Khomeinism—A Danger for Turkey," in David Menashri (ed.), The Iranian Revolution and the Muslim World (Boulder, CO: Westview Press, 1990), pp. 244–245.

[3] Unal Gundogan, Islamist Iran And Turkey 1979 1989…1979–1989: state pragmatism and ideological influences, *Middle East Review of International Affairs*, Vol. 7, No. 1 (March 2003).

[4] Sharq Newspaper, Turkey, was the most beneficiary of Iran-Iraq war, *Shargh Newspaper*, September 22, 2012, old.sharghdaily.ir/pdf/91-06-29/vijeh/13.pdf.

Ankara's burgeoning relations with Israel in the 1990s. The hardliners were greatly upset when, in 1996, Turkey signed the Military Training Agreement with Israel. The deal envisaged a range of joint training and information-sharing activities which gave Israel access to Turkish airspace. Other agreements provided for "technology transfer, joint research, intelligence-sharing, strategic planning talks, and bilateral and multilateral military exercises. One of them was the "Reliant Mermaid" search and rescue exercises, held in cooperation with the U.S. Navy in 1998 and 1999. Additionally, Israel received a contract to modernize 54 Turkish F-4s worth $650 million, a deal to upgrade 48 F-5s, and co-produce the Israeli Popeye II air-to-ground missile system. There were reported negotiations regarding Turkish participation in the Arrow antiballistic missile (ABM) program, co-production of Merkava tanks, and upgrades to Turkey's aging M-60 tanks. Although Turkish officials have asserted that the agreement was not a military pact, Tehran felt under threat. For the regime, which loathed the Zionist enemy, these extensive military ties brought Israel to its borders.[5]

Two further events exacerbated this tension. On February 15, 1999, Abdullah Öcalan, the fugitive head of the Turkish Kurdish Workers Party (PKK), was captured in Kenya with the alleged help of the U.S. Central Intelligence Agency (CIA) and the Israeli Mossad. This development which seemed to propel Turkey into a tighter bond with the United States was followed in April by Ankara's decision to move the American B57 tactical weapons to the Incirlik air base, near Adana. Some analysts, reflecting the Revolutionary Guards, argued that the deployment of the tactical weapons on Turkish soil made Turkey a "nuclear-weapon state." They contended that this should spur the Iranian leadership to invest in more ambitious nuclear capabilities—a code name for a nuclear arsenal.[6]

But the two countries' shared interests in containing the independent Kurdish movements mitigated the bad feelings. The PKK and the Iranian independence-seeking militants posed a threat to both countries. Collaboration dates back to 1988 when the two countries established the Turkey–Iran Joint Security Commission in 1988, followed by a Joint Security Protocol in November 1993, and the High-Security Commission

[5] Altunisik, "Turkish Policy Toward Israel," in Makovsky and Sayari, Turkey's New World: Changing Dynamics in Turkish Foreign Policy, p. 67.

[6] Mustafa Kibaroglu, Reassessing the Role of U.S. Nuclear Weapons in Turkey, June 4, 2010, https://www.armscontrol.org/act/2010_06/Kibaroglu.

in 1998. Iran's Interior Minister Mohammad Ali Besharati stated that his country would cooperate with Turkey in every way against their "common enemies." On his part, the Turkish Interior Minister Nahit Menteşe stated that Turkey would not allow any group operating from Turkish territory against Iranian interests. Rhetoric aside, the security protocols did not work well before the beginning of the twenty-first century. Turkish officials accused the Iranians of training violent Turkish Islamists and providing a safe haven for fleeing PKK fighters.[7]

In 2002, the Islamist Justice and Development Party (AKP) won the election, easing concerns in Tehran about the hyperactive secularist policy of the Kemalists—a group of political elites loyal to Mustafa Kemal Atatürk's reforms and ideologies. President Khatami praised the victory of Recep Tayyip Erdoğan and the Iran–Turkey Parliamentary Friendship Group Chairman, Ali Akbar Aghaee, declared that Iran was looking to improve its ties with Turkey. Aghaee explained that under the AKP both countries would work toward stability and peace in the region.[8]

The Turkish officials responded much more favorably. On March 8, 2002, the then General Secretary of the National Security Council (MGK), Tuncer Kılınc, stated that "Turkey would be better off trying to form alliances with Russia and Iran." The American invasion of Iraq in March 2003 brought Ankara and Tehran closer. Turkey refused the U.S. request to transit its forces through the Turkish port of Iskenderun into Iraq and banned the Americans from using the Incirlik base for bombing raids. The relations reached a nadir when the Americans seized 11 commandos from Turkish Special Forces in Northern Iraq on July 6, 2002. Kurdish sources maintained that Turkey sent the commandos to destabilize Northern Iraq but, according to another version, Ankara was worried that

[7] Gareth Jenkins, "Turkey and Iran Expected to Boost Security Cooperation," *Jamestown Foundation*, April 14, 2008, https://jamestown.org/program/turkey-and-iran-expected-to-boost-security-cooperation/; Olson, "The Kurdish Question and Turkey's Foreign Policy, 1991–1995; from Gulf War to Incursion into Iraq", *Journal of South Asian and Middle Eastern Studies*, Vol. 19, No. 1 (1995), p. 5; Seyyed Asadollah Athari, "Bazkhanie Ravaabete Do Keshvar pas az Jange Sard; Iran ve Torkeye", *Rahbord*, No. 27 (Spring 2003), pp. 268–269; Nader Entessar, "Kurdish Conflict in a Regional Perspective", in M.E. Ahrari, ed. Change and Continuity in the Middle East, Conflict Resolution and Prospects for Peace (New York: St. Martin's Press, 1996), pp. 47–73; TT, "From Now On Turkey not to Allow Anti-Iran Activities on Its Soil-16-Point MoU Signed between Iran, Turkey", *Tehran Times*, December 10, 1998.

[8] Hurriyet Daily, WORLD – Iran welcomes AKP government, *Hürriyet Daily News*, November 7, 2002; Erdogan Victory, A new Opportunity for Iran, Iran Diplomacy, www.irdiplomacy.ir/fa/page/13721/پیروزی+اردوغان،فرصتی+دوباره+برای+ایران.html.

the close relations between the Iraqi Kurds and the Americans would embolden the PKK.[9]

In 2004, the PKK and the newly reconfigured Partiya Jiyana Azad (PJAK) known as The Kurdistan Free Life Party—both based in the Qandil mountains of northern Iraq—resumed their respective campaigns. Finding themselves under attack, Tehran and Ankara revived their security collaboration. On July 29, 2004, Iran signed a Memorandum of Understanding with Ankara and soon after it declared the PKK to be a terror organization. For their part, the Turks helped Iran to hunt for PJAK fighters who were inflicting losses on the Iranian forces. Amid rumors that the United States and Israel were helping both the PKK and the PJAK, General Ilker Basburg, the commander of the Turkish land forces, stated that "Iran and Turkey have been conducting coordinated simultaneous operations on their borders." In February 2006 and April 2008, senior officials met at part of the High Commission deliberations.[10]

In 2008, Abbas Mohtaj, the Iranian deputy interior minister of Iran, and his Turkish counterpart Osman Gunes led their respective delegations. The two-day talks, which focused on the fight against terrorism and drugs, arms, and human trafficking, bore more fruit when the Islamic Revolutionary Guards Corps (IRGC) transferred 150 PKK militants to Turkey. Turkish media reported that militants had been caught after the Guards shelled PKK's camps located in the Qandil Mountains.[11]

[9] Sabah, Türkiye, Rusya ve İran'la ittifak arayışında olmalı, March 8, 2002, arsiv.sabah.com.tr/2002/03/08/p01.html; CNN, Turkey rejects U.S. troop proposal, March 2, 2003, edition.cnn.com/2003/WORLD/meast/03/01/sprj.irq.main/; Michael Howard and Susanne Goldenberg, "US Arrest of Soldiers Infuriates Turkey," *Guardian*, July 7, 2003; Latimes, Turkey Rejects U.S. Troop Deployment, *Los Angeles Times*, March 02, 2003, articles.latimes.com/2003/mar/02/world/fg-iraq2.

[10] Gareth Jenkins, "Turkey and Iran Expected to Boost Security Cooperation," *Jamestown Foundation*, April 14, 2008.

[11] Quoted in Suleyman Elik, Iran-Turkey Relations 1979–2011. Conceptualizing the Dynamics of Politics, p. 92, https://jamestown.org/program/turkey-and-iran-expected-to-boost-security-cooperation/; Dunya, Iran's deputy interior minister to visit Ankara, April 14, 2008, https://www.dunya.com/ekonomi/iran039s-deputy-interior-minister-to-visit-ankara-haberi-36148.

Beyond the Kurds: Working Out the Kinks in the Bilateral Relation

While the Kurdish problem was always at the forefront of Turkey's foreign policy, other factors played a role in the blossoming relations between the two countries. The hardliners were especially encouraged when the Prime Minister Recep Tayyip Erdoğan relaxed the stringent limitation on the display of religious symbols in the public domain. Tehran was particularly happy to see Ankara assume a more independent posture vis-à-vis Washington. By then both countries were middle powers, resisting the U.S. efforts to offshore balance regional players against each other, a posture which deflected pressure from Tehran.[12]

Other aspects of Erdoğan's foreign policy played into the blossoming relationship. Ankara adopted what IR theory calls "virtuous dynamics" to improve its regional stature. Accordingly, it promoted "zero problems" with its neighbors and offered mediation services to quell tensions and conflicts for a "sustainable modus vivendi" in the region. Turkey hoped to increase "interdependence by using political, social and economic channels, especially through private enterprise and the oil and natural gas pipelines network."[13]

In 2004 Prime Minister Erdoğan visited Tehran to sign several economic agreements, bolstering the burgeoning trade which climbed from $1 billion in 2000 to $4.3 billion in 2005. In 2011, trade reached $11 billion, making Turkey Iran's fifth largest trading partner. Cross-border trade between the two countries formed the second parameter in the bilateral relations. Turkish exports to Iran recorded an approximately tenfold increase from its level of around $350 million in 2002 to more than $3 billion in 2010. Turkish imports from Iran, which stood at around $850 million in 2002 reached $7 billion in 2010. Turkey's national statistics office released a report showing a 72.54 percent increase in Iran and Turkey's trade volume in 2011, compared to 2010. The report revealed that Iran and Turkey had a trade turnover of $7,762,997,000 during the first half of 2011.[14]

[12] Aaron Stein and Phillip C. Bleek, "Turkish-Iranian Relations: From Friends with Benefits to It is Complicated," *Insight Turkey*, Fall 2012.

[13] Anoushiravan Ehteshami ans Süleyman Elik, Turkey's Growing Relations with Iran and Arab Middle East, *Turkish Studies*, Vol. 12, No. 4, 2011.

[14] MNA, Iran-Turkey Trade Volume Surpassed $7.7 Billion, *Moj News Agency*, August 3, 2011.

Iran was particularly encouraged by Turkey's help in navigating its all-consuming nuclear problems. Chap. 2 noted that Tehran found itself increasingly isolated because of the Principalists, led by the high-profile President Ahmadinejad, created a move to refer the nuclear dossier to the United Nations Security Council (UNSC). The crippling sanctions had buffeted the Iranian economy, eroding the standards of living, and creating a legitimacy crisis for the regime. Turkey was one of the few countries willing to defy the United States and its allies. Ankara stressed that Iran had the right to enrich uranium and that the West could not take this right away as long as the enrichment process was peaceful as defined by the Non-Proliferation Treaty (NPT). President Erdoğan repeatedly vouched for Iran's compliance and peaceful intentions in the United Nations and the International Atomic Energy Agency (IAEA), earning him the sobriquet of "Iran's lawyer" in Washington.[15]

To prove that direct diplomatic engagement could work, Turkey, together with Brazil, negotiated an agreement with Ahmadinejad who was seeking ways to diffuse his standoff with the P5+1 group (the five permanent members of the UNSC and Germany). In 2010, both countries presented a deal to address concerns over Iranian production of high-enriched uranium (HEU). Turkey agreed to receive the stockpile of 1,200 kg of Low Enriched Uranium (LEU) from Iran in exchange for providing it with 120 kg of uranium enriched to 20 percent for the Tehran Research Reactor (TRR). The P5+1 was unpersuaded, and the deal was abandoned in favor of sanctions.[16]

Less public but equally important was Turkey's help in skirting the nuclear sanctions. Iran transferred a considerable amount of capital, previously held in Dubai, to Turkish banks, which were less prone to American pressure. Many Iranian businesspeople established themselves in Istanbul. Iranian investments in Turkey reportedly reached some $110 million in 2010. Turkish officials revealed that Iranians held deposits worth nearly $3 billion in Turkish banks. By some accounts, the Turkish deal helped save the Iranian economy from total collapse. Turkish tourism also profited from the international isolation of Tehran. The number of Iranian tourists visiting Turkey increased from around 330,000 in 2001 to over 1.8 million in 2010. This led to the establishment of new flight destinations.

[15] Aaron Stein and Phillip C. Bleek, "Turkish-Iranian Relations: From Friends with Benefits to It is Complicated, *Insight Turkey*, Fall 2012.

[16] ICG, In Heavy Waters: Iran's Nuclear Program, the Risk of War and the Lessons from Turkey," *International Crisis Group*, Report No. 116, February 23, 2012; Aaron Stein and Phillip C. Bleek, "Turkish-Iranian Relations: From Friends with Benefits to It is Complicated, *Insight Turkey*, Fall 2012.

The two neighbors also modernized existing border crossings, namely Gürbulak/Bazergan, Kapiköy/Razi and Esendere/Sero, with plans to open a new one at Dilucu/Maku.[17]

But even with the fortuitous convergence of interests, occasional disagreement marred the relations. Turkish dependence on Iranian oil and gas generated some ill feelings. In 1996, the two countries signed a deal to purchase natural gas over a twenty-year period. According to the contract, Iran was to deliver four billion cubic metres (BCM) in 2002 to rise to six BCM (the capacity of the pipeline) in 2007. But on January 19, 2006, Iran cut off supplies to Turkey due to frigid weather in Tabriz. In September 2006 and August 2007, supplies were interrupted again when three explosions, most likely caused by the PKK, hit the pipeline.[18]

Despite these occasional frictions, on April 16, 2014, the Generalized System of Preferences between Iran and Turkey was reached. It was implemented on January 1, 2015, paving the way for a mutual reduction in tariffs. Turkey was allowed to export 140 items on which tariffs were reduced by 20 to 50 percent, and Iran could export 125 products with a similar tariff reduction. Iranian critics noted that the agreement was skewed in Turkey's favor, but the government argued that there was an overlying rationale for maintaining good relations with Ankara.[19] Indeed, both countries were hoping to extend the friendship phase well into the future. But the Arab Spring, however, brought a sudden change to these plans.

The Arab Spring: Enemies More Than Friends

Like Iran, Turkey watched with interest the series of popular upheavals in 2011 but reached a very different conclusion about their meaning. In Tehran, the Supreme Leader Ayatollah Ali Khamenei asserted that "Iran's anti-Western

[17] Farhad Rezaei, Iran's Nuclear Program: A Study in Proliferation and Rollback (New York: Palgrave Macmillan, 2016).

[18] Manouchehr Moradi. Caspian Pipeline Politics and Iran-EU Relations. UNISCI Discussion Papers, 2006, p. 181; European Gas Markets: Heren Energy. Issues (news briefs): 15/09/2006, 29/09/2006, 17/09/2007.

[19] AISI, the preferential trade agreement between Iran and Turkey was signed, April 16, 2014, www.aisiran.com/node/1085; We Gave Rolling Pearl and Took Whey, January 6, 2015, https://www.tasnimnews.com/fa/news/1393/10/16/610481/درّ-غلتان-دادیم-و-آب-گرفتیم-کشک; Review of preferential trade agreement between Iran and Turkey, rc.majlis.ir/fa/report/show/955255.

and revolutionary example had borne fruit"—his code phrase for a new round of revolutionary export, as discussed in Chap. 1.[20]

But Ankara formed a different view of the events: "the Arab demand for electoral democracy was inspired by the example they had set in Turkey." In Erdoğan's view, it was Turkey's mission to help the Arab countries to build Islamist democracies. With the two visions on a collision course, the "Arab spring ushered in a winter for Turkish–Iranian relations," as one observer noted.[21]

Ankara was hopeful that several of the Arab Spring countries, including Egypt, Morocco, Algeria, and Tunisia, would base their domestic politics on the Turkish model. Indeed, Erdoğan counted on Muslim Brotherhood movements, which he considered an Arab equivalent of the AKP, to assume power in the Middle East. The Turkish leader was especially hopeful about the prospects of Mohammed Morsi, Egypt's Muslim Brotherhood chief, with whom he was very close. But the Iranian traditionalists were suspicious that Erdoğan was trying to utilize the unrest in the region to revive the Ottoman Empire. Critics accused him of being a neo-Ottoman leader and a modern-day Pharaoh on a quest to rule over the Sunni Arab countries. They contended that he was caught up in illusions of superiority and dreams of neo-Ottomanism.[22]

Ankara saw the popular protest in Syria as a good opportunity to test its political model. Turkish emissaries demanded that Assad launch democratic reforms to bring Syria in line with the spirit of the Arab Spring. Although in private, Assad promised to accommodate these demands, his brutal suppression of the protestors, some of whom fled to Turkey, angered Erdoğan. In June 2011, the Prime Minister stated that "They [Syrians] are not acting humanely. This is savagery." In August, the Turkish foreign

[20] Ayatollah Ali Khamenei, "Bayanat dar Didare Masoulan Nezam va Mihmanane Konferans Vahdat-e Eslami" Comments upon Visiting Officials of the System and Guests of the Islamic Unity Conference, speech given on January 19, 2014, http://farsi.khamenei.ir/speech-content?id=25056.

[21] Mustafa Gurbuz, "Turkey-Iran Relations: A Sectarian Future," *Arab Center, Washington*, March 8, 2017; Merv Tahiroglou and Behnam Ben Taleblu, "Turkey and Iran. The Best of Frenemies, *Turkish Policy Quarterly*, Vol. 14, No. 3, Spring 2015, 123–134.

[22] Payam Aftab, Erdogan and Abubakr Albaghdadi: Two Sides of one Coin, *Payam Aftab News Network*,www.payam-aftab.com/fa/doc/news/30034/سکه-يك-روي-دو-البغدادي-ابوبكر-اردوغان.

TNA, Erdogan Neo-Ottumanism or Democratization, *Tasnim News Agency*, April 18, 2017, https://www.tasnimnews.com/fa/news/1396/01/29/1382619/درّ-غلتان-داديم-و-آب-کشک-گرفتيم.

minister Ahmet Davutoglu went to Damascus to deliver a "tough message" while Erdoğan repeatedly complained about Syrian "savagery."[23]

Turkey suspended all diplomatic and economic ties with Syria and, according to sources, in August 2012, Ankara started secret talks with Washington on ways to remove Assad. On its own, the Turks launched a broad program to arm anti-Assad insurgents. They worked primarily with the Free Syrian Army (FSA), a coalition of Islamist groups formed on July 29, 2011. Colonel Riad Asaad along with six other defectors from Assad's army, including Lieutenant Ahmed Khalaf from Homs, and Captain Qais Qataneh, from Dara, headed the FSA. At various points, the group comprised the al-Sham Legion, the Levant Front, the Noureddine al-Zanki Battalions, the Jaish al-Tahrir and the Ahrar al-Sham.[24]

Unsurprisingly, the situation ratcheted up tensions between Turkey and Iran. Both sides blamed each other in "highly sectarian language," a discourse which one observer described as "unprecedented development in Turkish–Iranian relations." Turkey stressed that Iran's mobilization of Shiite militias from across the region to protect the rule of a minority sect, the Alawites, over a majority Sunni population in Syria had stirred up sectarian tensions across the region and provided Sunni jihadists with a potent recruitment tool. Ankara also noted that "contrary to conventional wisdom, Syria is not an Alawite (an offshoot of Shiism) regime," but a country where the Alawites oppressed the Sunni majority.[25]

Hardliners in Tehran responded in kind. Hassan Abbassi, the head of the Center for the Borderless Security Doctrinal Analysis noted "We must point out that we will not leave the Syrian people and its government alone against the allies of the Zionist regime. Our officials reminded Erdoğan when he visited Iran that if the Islamic Republic is forced to choose between Syria and Turkey, Iran will back Syria."[26]

[23] DB, "Turkish PM warns Syria against second Hama massacre". *Dünya Bülteni*, May 2, 2011; Milliyet, Şam'da kritik görüşme, August 9, 2011, www.milliyet.com.tr/sam-da-kritik-gorusme-siyaset-1424501/; NTV, Esad-Davutoğlu baş başa 3 saat görüştüler 9 Ağustos 2011, https://www.ntv.com.tr/turkiye/esad-davutoglu-bas-basa-3-saat-gorustuler,7gVzVHW7Gky1QGgCEBx_9Q.

[24] Declaration of the formation of the Free Syrian Army, *YouTube*, July 29, 2011, https://www.youtube.com/watch?v=SZcCbIPM37w.

[25] Middle East Report N°128, Syria's Mutating Conflict, 1 August 2012. Mustafa Gurbuz, Turkey-Iran Relation: A Sectarian Future," *Arab Center, Washington*, March 8, 2017; Crisis Group interviews, Turkish diplomats and security officials, Ankara, April, June 2016.

[26] Hassan Abbassi, Syria: Islamic World's Strategic Depth, in Mohammad Khalilpour, A Different Story of Logic of Being in Syria (Tehran: Masir-e Aftab, 2017).

Hossein Jaber Ansari, Iran's foreign ministry spokesman, warned Turkey that the continuation of policies and positions that "willfully or unwillingly lead to supporting terrorism in Syria and Iraq" will exacerbate the crisis in the region. Subsequently, the government spokesman, Mohammad Bagher Nobakht, told reporters that "if Erdoğan threatened Iran, he must know that the Foreign Minister of Iran said in a famous remark that one should never threaten an Iranian. Even stronger than Erdoğan could not threaten us." Ali Akbar Velayati, the foreign policy adviser to Khamenei, called on the Turkish authorities to pay respect, but the war of words had continued. Hassan Lasjerdi, Secretary of the Political Committee of the Islamic Coalition Party accused Turkey of supporting terrorist groups and Salafis, and Pirouz Mojtahedzadeh, a prominent Iranian political scientist called Erdoğan a "psycho" and a "madman".[27]

Mehr News Agency, an outlet that belongs to the Revolutionary Guards, wrote that Erdoğan and Davutoglu were in cahoots with the Saudis and Israelis. The agency noted "The behavior of the Turkish government is to serve Zionism and imperialism rather than the national interest of Turkey. In Iraq, there is no sectarian war between Shiites and Sunnis. But there is a war between the terrorists and the people and the Iraqi government, but those who are in the service of imperialism want to initiate the Shiite and Sunni war."[28]

Turkish support for the FSA elicited a more negative reaction from Tehran. Hossein Naqavi, the spokesman for the Majlis Foreign Affairs Committee, stated that "Turkey should keep out of Syria when it was unable to deal with its own internal affairs." Directing his remarks to Erdogan, Ahmad Reza Dastgheib, a Majlis member, urged Turkey and Saudi Arabia to abandon their support [of terrorists] as soon as possible.[29]

The weekly *Sobh Sadegh*, a magazine belonging to the IRGC, published an article titled "The Serious Stance of Iran Regarding the Syrian Crisis." The

[27] Realities, Accusations and Assumptions in Rhetorics Between Iran and Turkey, *Iran Diplomacy*, December 13, 2015, diplomacy.ir/fa/page/1954680/واقعیات،+اتهامات+و+فرضیهها+در+تنش+لفظی+ایران+و+ترکیه.html.

[28] MNA, Turkey Should Not Give its Steer to Saudis, *Mehr News Agency*, January 1, 2016, www.mehrnews.com/news/3011963/ترکیه-نباید-سکان-امور-خود-را-به-آل-سعود-بسپارد.

[29] Quoted in Stephen Larabee and Alireza Nader, Turkish-Iranian Relations in changing Middle East (Santa Monica: Rand Corporation, 2011); Turkey and Saudi Interference in Syrian Internal Affairs is a Big Mistake, *YJC*, February 12, 2016, www.yjc.ir/fa/news/5494918/دخالتهای-عربستان-و-ترکیه-در-امور-سوریهاشتباه-بزرگی-است.

article stated: "When the authorities of Turkey continue to insist on their position towards Syria, which leads to further tensions, it urges Iran to choose one between Turkey and Syria. The logic of strategic interests will lead Iran to choose Syria. Is it good that the Turkish authorities understand this Iranian necessity and do not let Iran reach that point."[30]

When President Rouhani came to power in 2103, the tone and policy in Tehran changed. As Chap. 5 described, the moderates considered the Syrian engagement to be an expensive overreach and one that threatened their normalization project. In September 2013, Rouhani floated a peace plan which the Foreign Minister Javad Zarif took to Ahmet Davutoglu, his counterpart in Ankara. After months of negotiations, the plan failed because of disagreements about the future of Assad. The Turks wanted Assad barred from running in a U.N.-monitored election, but the Iranian balked, arguing that he was not popular enough to win an election in the first place. But their interlocutors were persuaded that Assad would manipulate the ballot and decided to wait for his military defeat. According to the then Turkish president Abdullah Gul, "our government did not pursue an agreement with Iran because it thought Assad would be toppled in a few months."[31]

Albeit a significant issue, Syria was not the only irritant in the relations between the two countries. The Turkish government was also worried about the inroads that the Iranians had made in Iraq. Officials in Ankara charged the IRGC-QF with persecuting the Sunnis and fomenting ethnic strife. Erdoğan went as far as to accuse Iran of seeking to resurrect the Persian Empire and impose Shiite domination to parts that were historically Ottoman. In March 2015, the Turkish leader asserted that "Iran was fighting IS in Iraq only to take its place." He added that Turkey "cannot tolerate Iran's bid to dominate the Middle East."[32]

But it was Iran's decision to support the Patriotic Union of Kurdistan (PUK) led by Jalal Talabani, which Ankara considered to be an affiliate of the PKK, that was particularly explosive. Ankara, which supported the

[30] SSW, Between Turkey and Syria, Iran will Choose Assad, *Sobhe Sadegh Weekly*, July 18, 2011, www.asriran.com/fa/news/173230/ایران-سپاه-ارگان-میان-ترکیه-و-سوریه-اسد-را-انتخاب-می-کند.

[31] ICG, Turkey and Iran: Bitter Friends, Bosom Rivals, *International Crisis Group*, https://www.crisisgroup.org/middle-east-north-africa/gulf-and-arabian-peninsula/iran/b051-turkey-and-iran-bitter-friends-bosom-rivals.

[32] Crisis Group interviews, Turkish officials, Ankara and Istanbul, March–August 2016. Humeyar Pamuk, "Turkey's Erdoğan says can't tolerate Iran bid to dominate Middle East" *Reuters*, March 26, 2015.

Kurdish Democratic Party (KDP) under Moussa Masoud Barazani, invested heavily in the relations with the Kurdish Provisional Government (KPG).

Certain regional issues proved equally divisive. In Yemen, Turkey backed Saudi Arabia whereas Iran backed the Houthis as noted in the preceding chapter. The policy disagreement led to a very public spat between the two capitals. Erdoğan declared that "Iran is trying to dominate the region…This has begun annoying us, Saudi Arabia, and the Gulf countries. This is really not tolerable and Iran has to see this." He went on to say that "Iran and the terrorist groups must withdraw" from Yemen.[33] Foreign Minister Zarif responded that "Turkey makes strategic mistakes." However, as befitting the frenemies relationships, a few days later, Erdogan arrived in Tehran for discussions on improving Turkish–Iranian trade relations where Khamenei and Rouhani had received him.[34]

This interlude did not prevent hardline media in Tehran from attacking Turkey. An article in the Fars News Agency associated with the Revolutionary Guards declared that Ankara is a "traditional ISIS supporter" and, together with Saudi Arabia has "entered Yemen," where it is repeating its success of "massacring the Syrian." The article concluded that through the triangle of "Ankara, Doha, Riyadh, Erdogan is trying to once again give new breath to the dead body of ISIS and Takfiri groups in the region."[35]

Operation Euphrates Shield: The View from Ankara and Tehran

By late 2015, observers in Ankara were increasingly worried about the developments in Syria. Expectations that the Turkish-backed FSA would defeat Assad were replaced by anxiety over the Syrian Kurds. Two groups working with the United States, the Kurdish Syrian Democratic Forces (SDF) and the People's Protection Units (YPG), was making strides

[33] Haberler, Iran and the terrorist groups must withdraw (Cumhurbaşkanı Erdoğan: İran ve Terörist Gruplar Yemen'den Çekilmeli), March 26, 2015, https://www.haberler.com/cumhurbaskani-erdogan-france-24-e-konustu-iran-ve-7123575-haberi/.

[34] FNA, Zarif Reaction to Erdogan Statements, *Fars News Agency*, March 27, 2015, www.farsnews.com/newstext.php?nn=13940107000307.

[35] FNA, Turkey and Saudi Arabia Ominous Plan for Yemen, *Fars News Agency*, Mar 21, 2015, af.farsnews.com/world/news/13940101000394; MNA, Turkey-Saudi Arabia New Sedition is Forthcoming: Meeting of ISIS Supporters in Riyadh, *Mehr News Agency*, December 20, 2015, www.mehrnews.com/news/3005233/فتنه-جدید-ترکیه-و-سعودی-در-راه-است-دیدار-حامیان-داعش-در-ریاض.

against ISIS and were on the way to create a Kurdish contiguity along the 510-mile border which Turkey shared with Syria. To undermine this prospect, Ankara launched Operation Euphrates Shield. A short communique of August 24, 2016, expressed its desire to topple the "dictator Bashar Assad." The Iranian response to Turkey's military incursion was swift. Bahram Qassem, the Foreign Ministry spokesman, urged all countries to respect Syrian's national sovereignty and territorial integrity: "Although the fight against terrorism and efforts to maintain regional stability and security are an unchangeable and important principle in the foreign policy of peace-seeking countries in the region, this issue cannot and should not be used as a justification for violating the territorial integrity of another country by conducting military operations against that country without coordination with its central government, and by overlooking its national sovereignty."[36]

Diplomatic talk aside, the IRGC-QF understood that Turkey was anxious to prevent the consolidation of the Democratic Federation of Northern Syria known as Rojava. A de facto entity since 2012, Rojava contained three cantons—Afrin Canton, Jazira Canton, and Kobani Canton—and the Shahba region. The Iranian media quoted Mehmet İlker Başbuğ, a former Turkish Chief of Staff, to the effect that the formation of a Kurdish corridor in the Turkish border would pose a grave threat to Turkey.[37]

But Iran expressed "concern" about the operation of the Turkish army in Syria, stating that Turkey's entry into Jerablus is an aggression without international consent in Damascus.[38] Brigadier General Yadullah Javani, the head of the Guards political bureau and Ayatollah Khamenei's military adviser, added: "Although Turkey has declared that it has joined the anti-ISIS coalition, the behavior of the Turkish government shows that their main goal is not confronting ISIS. Erdoğan's remarks that Turkey's top priority is the fall of Bashar al-Assad proves that fighting ISIS is not Turkey's concern." Javani went on by saying that "not to mention that

[36] Global Security, Operation Euphrates Shield, https://www.globalsecurity.org/military/world/war/syria-euphrates-shield.htm.

[37] Turkey is Worried About Syrian Kurds Autonomy, *Iranian Diplomacy*, June 10, 2016, www.irdiplomacy.ir/fa/page/1959934/ترکیه+نگران+از+خودمختاری+کردهای+سوریه.html.

General Suleimani Could Read Turkey's Hand, *Serat News Agency*, August 5, 2016, www.seratnews.ir/fa/news/314703/وقتی-سردار-سلیمانی-دست-ترکیه-را-خواند-ماجرای-عصبانیت-اردوغان-در-دیدار-با-لاریجانی.

[38] Dolate Bahar, The "Euphrates Shield" operation was the West's favor to Erdoğan, August 29, 2016, www.dolatebahar.com/telex-303645.html; Press TV, Iran expressed "concern" over the Turkish military operation in Syria, August 31, 2016, parstv.tv/ایران-از-ادامه-عملیات-ارتش-ترکیه-در-سور/.

Ankara played a significant coordinating role in the formation of terrorist groups, especially the ISIS. Terrorists were trained in military camps in Turkey and Turks have treated the wounded terrorists and provided them with training and equipment."[39]

Luckily for QF, Euphrates Shield did not go according to plan. Although the Turkish forces captured several towns in the area, including al-Bab, Jarablus, and Dabiq, the U.S. moved its special operation forces from south of the Sajuk River to north of Manbij to protect the SDF fighters. Russian military personnel which monitored the cease-fire near Afrin extended the same kind of protection to the YPG units. Boxed in the east and the west, the Turks declared the end of Operation Euphrates Shield in March 2017.[40]

If the American and Russian intervention spared Tehran direct clashes with Ankara, the situation in Iraq was more complicated. With the battle for Mosul nearing its end, Turkey was anxious to gain a political foothold in post-ISIS Iraq. The opportunity presented itself when ostensibly responding to an invitation by Barzani; Turkey sent an expeditionary force of 2000 to the Bashiqa camp in northern Iraq in October 2016. Among other tasks, the Turkish troops were to watch out for efforts by the Shiite militias to change the demography of Mosul, a practice which the QF had instituted in territories liberated from ISIS in Syria.[41]

The Iraqi government of Prime Minister Heider al-Abadi reacted strongly against the Turkish incursion. To recall Chap. 5, al-Abadi was pressured to incorporate the Iranian-supported Popular Mobilization Forces (PMF) into the Iraqi military. But the Sunnis, still reeling from the persecution by the Shiite militias in the mid-2000s, were apprehensive about the PMF, a cause that President Erdoğan took up. He called the PMF a "terrorist organiza-

[39] Yadullah Javani, Iran's Fundamental and Sincere Defending of Regional Nations, in Mohammad Khalilpour, A Different Story of Logic of Being in Syria (Tehran: Masir-e Aftab, 2017), pp. 103–109.

[40] Turkey can start new operation if necessary as Euphrates Shield ends: PM, *Hurriyet Daily*, March 29, 2017, www.hurriyetdailynews.com/turkeys-euphrates-shield-operation-in-syria-successfully-completed.aspx?pageID=238&nID=111401&NewsCatID=352; Cenzig Candar, "Operation Euphrates Shield: A Postmortem," *Al-Monitor*, April 5, 2017; Iran welcomed the end of the operation but called on Turkey to end its involvement in Syria and to withdraw its forces from Syria immediately. Iran Called on Turkey to Withdraw from Syria, *Afkar News*, www.afkarnews.ir/بخش-سیاسی-3/585571-درخواست-ترکیه-از-ایران-برای-خروج-از-سوریه.

[41] Birce Bora, "What is Turkey Trying to Achieve in Iraq," *Al Jazeera*, October 16, 2016; Turning Mosul into a Base Against Iran, Paydari Melli, December 6, 2015, paydarymelli.ir/fa/news/18822/تبدیل-موصل-به-پایگاهی-علیه-ایران.

tion" and criticized Iran's "expansionist" policies in the region. In return, the PMF urged al-Abadi to expel the Turkish ambassador from Baghdad and Karim al-Nuri a senior PMF commander told an IRGC media outlet that Turkey would be "isolated" from Iraq.[42]

With ISIS losing ground in Iraq, the tensions between the two frenemies grew. In February 2017, President Erdoğan accused Iran of resorting to Persian nationalism to split Iraq and Syria. His Foreign Minister Mevlut Cavasoglu added that "Iran is trying to create two Shiite states in Syria and Iraq. This is very dangerous. It must be stopped." Iran responded by summoning the Turkish ambassador to warn him that the Iranian "patience has limits." Turkey reiterated that as a protector of the Sunnis, it is responsible for Iraq. But Iran warned that the Turkish "incursion" must stop and hinted that Kurdish forces which Turkey considers "terrorists" may take over the area occupied by the Turkish expeditionary force.[43]

To the uniformed observers, these public exchanges looked like a new low in the frenemies' relations. However, as in previous episodes, the hostile rhetoric masked a turn into a friendlier territory. Ironically, it was the situation in Syria which triggered the chain of events leading to this change.

The Frenemies Endgame: Iran–Turkish Space—Sharing in the Middle East?

To recall Chaps. 3 and 6, Russia, Iran, and Turkey reached the de-escalation zones agreement in Astana on May 4, 2017. The deal that served as a basis for the unofficial Tripartite Agreement gave Turkey, which failed to execute the Euphrates Shield operation, an unexpected victory.[44]

As part of the de-escalation deal, Russian and Turkish personnel were deployed to Idlib in northern Syria. Under Russian coverage, Turkey deployed some 500 soldiers close to the Kurdish enclave of Afrin, where

[42] Ahmad Majidar, "Iran-Backed Iraqi Mobilization Forces Urge Baghdad to Expel Turkish Ambassador," *Iran Observed, Middle East Institute*, April 24, 2017.

[43] Birce Bora, "What is Behind the Hostility between Iran and Turkey," *Al Jazeera*, February 26, 2017.

[44] Josie Ensor, Turkey, Russia and Iran sign 'safe zones' deal for Syria, *The Telegraph*, May 4, 2017, www.telegraph.co.uk/news/2017/05/04/turkey-russia-iran-sign-safe-zones-deal-syria/; Turkey, Iran, and Russia sign deal on safe zones in Syria, *The Nationale*, May 4, 2017, https://www.thenational.ae/world/turkey-iran-and-russia-sign-deal-on-safe-zones-in-syria-1.34497.

the U.S. supported Kurdish Democratic Union Party (PYD) was based. Erodgan's spokesman Ibrahim Kalin provided the Turkish view of the Astana de-escalation deal: "We will probably be most prominent in the Idlib region with the Russians; mostly Russia and Iran around Damascus, and a mechanism involving the Americans and Jordan in the south in the Deraa region is being worked on. Kurdish sources reported that the Turkish contingent shelled villages where YPG forces were based. The IRGC-QF denied reports of collusion with the Turkish force, but it certainly helped that both countries shared a hostility toward the Kurds. With Ankara finally ready to accept the Assad regime, the Erbil offensive was a positive in the frenemy calculus."[45]

The Saudi project to isolate Qatar described in the preceding chapter added more positives to the ledger. Historically, Turkey enjoyed close relations with Qatari leaders, not least because of their joint support for the Muslim Brotherhood and the Palestinian Hamas. Ankara's reaction to the Saudi action was swift. On June 7, 2017, the parliament ratified a military agreement with Qatar under which Turkish troops were deployed to a Turkish military base in Qatar.[46]

Both Iran and Turkey shipped several tons of food daily to Qatar to ease the blockade imposed by the Saudi coalition. But the food shipment represented more than a humanitarian gesture. Since Riyadh demanded that Qatar to reduce its diplomatic ties with Iran and close the Turkish base, the joint action was a resounding no to the Kingdom.[47]

[45] Reuters, Turkey and Russia to deploy in Syria's Idlib: Erdogan spokesman, *Reuters*, June 22, 2017, https://www.reuters.com/article/us-mideast-crisis-syria-turkey-russia/turkey-and-russia-to-deploy-in-syrias-idlib-erdogan-spokesman-idUSKBN19D1WI; Metin Gurcan, Ankara's offer to Moscow: Give us Afrin for Idlib, *Al-Monitor*, July 3, 2017, www.al-monitor.com/pulse/originals/2017/06/turkey-russia-offer-moscow-idlib-in-return-for-afrin.html.

[46] Sigurd Neubauer, The curious timing of the Qatar crisis, *Al-Monitor*, June 6, 2017, www.al-monitor.com/pulse/originals/2017/06/timing-qatar-crisis-saudi-arabia-gcc-uae-egypt-diplomatic.html; Aljazeera, Turkish parliament approves troop deployment to Qatar, June 7, 2017, www.aljazeera.com/news/2017/06/turkey-fast-track-troops-deployment-qatar-170607151127104.html; Metin Gurcan, Turkish military will protect Qatari regime if necessary, *Al-Monitor*, June 12, 2017, www.al-monitor.com/pulse/originals/2017/06/turkey-qatar-ankara-firm-stance-on-el-thani-keeps-his-seat.html.

[47] Reuters, Qatar in talks with Turkey and Iran to provide food, water: official, June 7, 2017,https://www.reuters.com/article/us-gulf-qatar-food/qatar-in-talks-with-turkey-and-iran-to-provide-food-water-official-idUSKBN18Y0Z8; Iran and Turkey Filled Saudi Place in Qatar Market, fa.euronews.com/2017/06/21/turkish-and-iranian-products-on-the-shelves-of-qatari-super-markets; Saeid Jafari, Saudi-led Qatar blockade brings Iran, Turkey

The Kurdish Iraqi referendum scheduled for September 25, 2017, however, was arguably the engine most instrumental in driving the frenemy equation into positive territory. As noted in Chap. 5, both countries were bitterly opposed to Kurdish independence out of fear that their Kurdish minorities would follow suit. A closer military coordination was a logical step toward combating the PKK and PJAC. This much became clear when the Iranian Chief of Staff Major General Mohammad Bagheri traveled to Ankara at the head of a large delegation on August 15, 2017. He met with his counterpart General Hulusi Akar, with President Erdoğan, and senior Turkish officials. According to Bagheri, the visit was "necessary to exchange views and more cooperation on the military subjects and different regional issues, issues related to the two countries' security, security of borders and fighting terrorism." Iran's Foreign Ministry Spokesman Bahram Qassemi told reporters: "The trip was exceptional in that we did not have such visits at such levels for the past 40 years. … The talks were important and decisive. Maj. Gen. Bagheri had a productive meeting with the Turkish president."[48]

A well-informed source in Tehran revealed that the meetings in Turkey were "very important to tackle issues that are eminent and have dire consequences on the national security of both countries. The referendum on [Iraqi] Kurdistan is a matter that concerns Tehran for several reasons. The same applies to Turkey; therefore, all options—literally all options—were on the table of discussions. Both countries are going to take necessary steps to make sure Iraqi Kurdistan doesn't split from Iraq, by any means necessary."[49]

President Erdoğan confirmed that the discussion included the referendum in Iraqi Kurdistan, the situation in the Idlib province, and the need to combat terrorism through a joint military campaign in northern Iraq against the PKK and PJAK. A source close to the president told the

together, July 3, 2017, www.al-monitor.com/pulse/originals/2017/06/saudi-qatar-blockade-iran-turkey-convergence-crisis.html; Erdogan, Arab Countries Violate International Law in Qatar Crisis, June 25, 2017, fa.euronews.com/2017/06/25/erdogan-says-arab-demands-on-qatar-are-against-international-law; Ali Hashem, al Monitor Iran, Turkey move to re-establish role as regional backbone, *Al-Monitor*, August 23, 2001, http://www.al-monitor.com/pulse/en/originals/2017/08/iran-turkey-iraqi-kurdistan-krg-syria-regional-order-bagheri.amp.html.

[48] Etemad, Iran-Turkey Entered New Phase, http://etemadnewspaper.ir/?News_Id=84035.

[49] Ali Hashem, Iran, Turkey move to re-establish role as regional backbone, *Al-Monitor*, August 23, 2017, www.al-monitor.com/pulse/originals/2017/08/iran-turkey-iraqi-kurdistan-krg-syria-regional-order-bagheri.html.

Al-Monitor publication that a joint military campaign was Erdoğan's initiative. In his view, "the PKK is a threat to Turkey's national security as well as to Iran's, while PJAK, which we are already fighting, is the one posing a threat to Iran's [national security]. He wanted to see both his country and Iran launching the campaign against both groups, and in return, he's ready to be more helpful in Syria. Idlib could be a step of goodwill; still, Iran doesn't see the PKK as an enemy."[50]

The Guards, however, denied plans to operate outside Iran: "We have not planned any operational programs outside the borders of the Islamic Republic of Iran, but as with the past, we will severely retaliate against each team or individual who intends to infiltrate the Islamic Republic of Iran for acts of terrorism or hostility."[51]

Still, some analysts questioned the extent of the Guards' involvement in the Bagheri visit. According to Amir Taheri, a long-time Iran observer, the IRGC-QF "was forced to take a back seat to the more mainstream Artesh [Iran's regular army] and, by extension, the Rouhani government." Taheri noted that the "understanding in Tehran was that Iran cannot achieve its goal in Syria and the Levant at large without considering the interests of the Turks." As for Ankara, it preferred the Artesh because of its suspicion that the Guards had manipulated the PKK and PUK against Turkish interests in Iraq.[52]

Even if the Rouhani government did gain the upper hand in dealing with Turkey, the Revolutionary Guards were dictating the policy about Israel.

Conclusion

Iran and Turkey have a long history of rivalry, which has hampered the more recent efforts to shape the region consistently with their respective visions. During the sanctions regime, Tehran was seeking a reliable trading partner and Ankara needed access to Iranian oil. Turkish private deals helped Iran prevent a total collapse of its economy. Even post-JCPOA, the trade relations have remained robust.

But the historical rivalry between Shiite Iran and under President Erdoğan, who considers himself the leader of the Sunnis in the region,

[50] Ibid.

[51] MNA, We Have no Operational Plan for Outside of Borders, *Mehr News Agency*, August 22, 2017, www.mehrnews.com/news/4066740/هیچ برنامه-عملیاتی-در-خارج-از-مرزها-برنامه-ریزی-نکرده-ایم.

[52] Amir Taheri, "Puffing the Turkish Chibouk in Ankara," *Asharq al Awsat*, August 24, 2017.

made it difficult to achieve more than a modest cooperation in the Syrian conflict. The future of the Assad regime is the most vexing in this respect as Erdoğan would want to see the regime fall.

The fight against ISIS in Iraq has opened another fissure between the two countries. Having established a strong influence in Iraq in the wake of the American invasion in 2003, Tehran has hoped to expand its reach around Mosul. To prevent the Shia dominance, however, Turkey has sent its own troops to areas bordering on Mosul, an action that Tehran strongly rebuked. Unlike in other foreign policy areas, there is virtually no difference between the normalizers and Principalists in dealing with Turkey. This rare consensus has also carried over to the current negotiations on the future of Syria in which both Iran and Turkey are represented, along with Russia.

References

Aaron Stein and Phillip C. Bleek, "Turkish-Iranian Relations: From Friends with Benefits to It is Complicated," *Insight Turkey*, Fall 2012.

Ahmad Majidar, "Iran-Backed Iraqi Mobilization Forces Urge Baghdad to Expel Turkish Ambassador," *Iran Observed, Middle East Institute*, April 24, 2017.

AISI, the preferential trade agreement between Iran and Turkey was signed, April 16, 2014, www.aisiran.com/node/1085.

Ali Hashem, al Monitor Iran, Turkey move to re-establish role as regional backbone, *Al-Monitor*, August 23, 2001, http://www.al-monitor.com/pulse/en/originals/2017/08/iran-turkey-iraqi-kurdistan-krg-syria-regional-order-bagheri.amp.html.

Ali Hashem, Iran, Turkey move to re-establish role as regional backbone, *Al-Monitor*, August 23, 2017, www.al-monitor.com/pulse/originals/2017/08/iran-turkey-iraqi-kurdistan-krg-syria-regional-order-bagheri.html.

Aljazeera, Turkish parliament approves troop deployment to Qatar, June 7, 2017, www.aljazeera.com/news/2017/06/turkey-fast-track-troops-deployment-qatar-170607151127104.html.

Altunisik, "Turkish Policy Toward Israel," in Makovsky and Sayari, Turkey's New World: Changing Dynamics in Turkish Foreign Policy.

Amir Taheri, "Puffing the Turkish Chibouk in Ankara," *Asharq al Awsat*, August 24, 2017.

Anoushiravan Ehteshami ans Süleyman Elik, Turkey's Growing Relations with Iran and Arab Middle East, *Turkish Studies*, Vol. 12, No. 4, 2011.

Ayatollah Ali Khamenei, "Bayanat dar Didare Masoulan Nezam va Mihmanane Konferans Vahdat-e Eslami" Comments upon Visiting Officials of the System and Guests of the Islamic Unity Conference, speech given on January 19, 2014, http://farsi.khamenei.ir/speech-content?id=25056.

Birce Bora, "What is Behind the Hostility between Iran and Turkey," *Al Jazeera*, February 26, 2017.

Cenzig Candar, "Operation Euphrates Shield: A Postmortem," *Al-Monitor*, April 5, 2017.

CNN, Turkey rejects U.S. troop proposal, March 2, 2003, edition.cnn.com/2003/WORLD/meast/03/01/sprj.irq.main/.

Crisis Group interviews, Turkish diplomats and security officials, Ankara, April, June 2016.

Crisis Group interviews, Turkish officials, Ankara and Istanbul, March–August 2016.

Crisis Group, Briefing No. 51, Middle East and North Africa, December 13, 2016.

DB, "Turkish PM warns Syria against second Hama massacre." *Dünya Bülteni*, May 2, 2011.

Declaration of the formation of the Free Syrian Army, *YouTube*, July 29, 2011, https://www.youtube.com/watch?v=SZcCbIPM37w.

Dolate Bahar, The "Euphrates Shield" operation was the West's favor to Erdoğan, August 29, 2016, www.dolatebahar.com/telex-303645.html.

Dunya, Iran's deputy interior minister to visit Ankara, April 14, 2008, https://www.dunya.com/ekonomi/iran039s-deputy-interior-minister-to-visit-ankara-haberi-36148.

Erdogan Victory, A new Opportunity for Iran, Iran Diplomacy, www.irdiplomacy.ir/fa/page/13721/پیروزی+اردوغان،فرصتی+دوباره+برای+ایران.html.

Erdogan, Arab Countries Violate International Law in Qatar Crisis, June 25, 2017, fa.euronews.com/2017/06/25/erdogan-says-arab-demands-on-qatar-are-against-international-law.

Ergun Ozbudun, "Khomeinism—A Danger for Turkey," in David Menashri (ed.), The Iranian Revolution and the Muslim World (Boulder, CO: Westview Press, 1990.

Etemad, Iran-Turkey Entered New Phase, http://etemadnewspaper.ir/?News_Id=84035.

European Gas Markets: Heren Energy. Issues (news briefs): 15/09/2006, 29/09/2006, 17/09/2007.

Farhad Rezaei, Iran's Nuclear Program: A Study in Proliferation and Rollback (New York: Palgrave Macmillan, 2016).

FNA, Turkey and Saudi Arabia Ominous Plan for Yemen, *Fars News Agency*, March 21, 2015, af.farsnews.com/world/news/13940101000394.

FNA, Zarif Reaction to Erdogan Statements, *Fars News Agency*, March 27, 2015, www.farsnews.com/newstext.php?nn=13940107000307.

Gareth Jenkins, "Turkey and Iran Expected to Boost Security Cooperation," *Jamestown Foundation*, April 14, 2008, https://jamestown.org/program/turkey-and-iran-expected-to-boost-security-cooperation/.

General Suleimani Could Read Turkey's Hand, *Serat News Agency*, August 5, 2016, www.seratnews.ir/fa/news/314703/ماجرای-خواند-را-ترکیه-دست-سلیمانی-سردار-وقتی-لاریجانی-با-دیدار-در-اردوغان-عصبانیت.

Global Security, Operation Euphrates Shield, https://www.globalsecurity.org/military/world/war/syria-euphrates-shield.htm.

Haberler, Iran and the terrorist groups must withdraw (Cumhurbaşkanı Erdoğan: İran ve Terörist Gruplar Yemen'den Çekilmeli), March 26, 2015, https://www.haberler.com/cumhurbaskani-erdogan-france-24-e-konustu-iran-ve-7123575-haberi/.

Hassan Abbassi, Syria: Islamic World's Strategic Depth, in Mohammad Khalilpour, A Different Story of Logic of Being in Syria (Tehran: Masir-e Aftab, 2017).

Humeyar Pamuk, "Turkey's Erdoğan says can't tolerate Iran bid to dominate Middle East" *Reuters*, 26 March 2015.

Hurriyet Daily, WORLD – Iran welcomes AKP government, *Hürriyet Daily News*, November 7, 2002.

ICG, "In Heavy Waters: Iran's Nuclear Program, the Risk of War and the Lessons from Turkey," *International Crisis Group*, Report No. 116, February 23, 2012.

ICG, Turkey and Iran: Bitter Friends, Bosom Rivals, *International Crisis Group*, https://www.crisisgroup.org/middle-east-north-africa/gulf-and-arabian-peninsula/iran/b051-turkey-and-iran-bitter-friends-bosom-rivals.

Iran and Turkey Filled Saudi Place in Qatar Market, fa.euronews.com/2017/06/21/turkish-and-iranian-products-on-the-shelves-of-qatari-supermarkets.

Iran Called on Turkey to Withdraw from Syria, *Afkar News*, www.afkarnews.ir/بخش-سیاسی-3/585571-درخواست-ایران-از-ترکیه-برای-خروج-از-سوریه.

Josie Ensor, Turkey, Russia and Iran sign 'safe zones' deal for Syria, *The Telegraph*, May 4, 2017, www.telegraph.co.uk/news/2017/05/04/turkey-russia-iran-sign-safe-zones-deal-syria/.

Latimes, Turkey Rejects U.S. Troop Deployment, *Los Angeles Times*, March 2, 2003, articles.latimes.com/2003/mar/02/world/fg-iraq2.

Manouchehr Moradi, Caspian Pipeline Politics and Iran-EU Relations. UNISCI Discussion Papers, 2006.

Merv Tahiroglou and Behnam Ben Taleblu, Turkey and Iran The Best of Frenemies, *Turkish Policy Quarterly*, Vol. 14, no 3, Spring 2015.

Metin Gurcan, Ankara's offer to Moscow: Give us Afrin for Idlib, *Al-Monitor*, July 3, 2017, www.al-monitor.com/pulse/originals/2017/06/turkey-russia-offer-moscow-idlib-in-return-for-afrin.html.

Metin Gurcan, Turkish military will protect Qatari regime if necessary, *Al-Monitor*, June 12, 2017, www.al-monitor.com/pulse/originals/2017/06/turkey-qatar-ankara-firm-stance-on-el-thani-keeps-his-seat.html.

Michael Howard and Susanne Goldenberg, "US Arrest of Soldiers Infuriates Turkey," *Guardian*, July 7, 2003.

Middle East Report N°128, Syria's Mutating Conflict, 1 August 2012.

Milliyet, Şam'da kritik görüşme, August 9, 2011, www.milliyet.com.tr/sam-da-kritik-gorusme-siyaset-1424501/.

MNA, Iran-Turkey Trade Volume Surpassed $7.7 Billion, *Moj News Agency*, August 3, 2011.

MNA, Turkey Should Not Give its Steer to Saudis, *Mehr News Agency*, January 1, 2016, www.mehrnews.com/news/3011963/ترکیه-نباید-سکان-امور-خود-را-به-آل-سعود-بسپارد.

MNA, Turkey-Saudi Arabia New Sedition is Forthcoming: Meeting of ISIS Supporters in Riyadh, *Mehr News Agency*, December 20, 2015, www.mehrnews.com/news/3005233/فتنه-جدید-ترکیه-و-سعودی-در-راه-است-دیدار-حامیان-داعش-در-ریاض.

MNA, We Have no Operational Plan for Outside of Borders, *Mehr News Agency*, August 22, 2017, www.mehrnews.com/news/4066740/هیچ-برنامه-عملیاتی-در-خارج-از-مرزها-برنامه-ریزی-نکرده-ایم.

Mustafa Gurbuz, "Turkey-Iran Relations: A Sectarian Future," *Arab Center, Washington*, March 8, 2017.

Mustafa Kibaroglu, Reassessing the Role of U.S. Nuclear Weapons in Turkey, June 4, 2010, https://www.armscontrol.org/act/2010_06/Kibaroglu.

Nader Entessar, "Kurdish Conflict in a Regional Perspective," in M.E. Ahrari, ed. Change and Continuity in the Middle East, Conflict Resolution and Prospects for Peace (New York: St. Martin's Press, 1996).

NTV, Esad-Davutoğlu baş başa 3 saat görüştüler 9 Ağustos 2011, https://www.ntv.com.tr/turkiye/esad-davutoglu-bas-basa-3-saat-gorustuler,7gVzVHW7Gky1QGgCEBx_9Q.

Olson, "The Kurdish Question and Turkey's Foreign Policy, 1991–1995: From Gulf War to Incursion into Iraq", *Journal of South Asian and Middle Eastern Studies*, Vol. 19, No. 1 (1995).

Payam Aftab, Erdogan and Abubakr Albaghdadi: Two Sides of one Coin, *Payam Aftab News Network*, www.payam-aftab.com/fa/doc/news/30034/اردوغان-ابوبكر-البغدادي-دو-روي-يك-سكه.

Press TV, Iran expressed "concern" over the Turkish military operation in Syria, August 31, 2016, parstv.tv/ایران-از-ادامه-عملیات-ارتش-ترکیه-در-سور/.

Realities, Accusations and Assumptions in Rhetorics Between Iran and Turkey, *Iran Diplomacy*, December 13, 2015, diplomacy.ir/fa/page/1954680/واقعیات،+اتهامات+و+فرضیهها+در+تنش+لفظی+ایران+و+ترکیه.html.

Reuters, Qatar in talks with Turkey and Iran to provide food, water: official, June 7, 2017, https://www.reuters.com/article/us-gulf-qatar-food/qatar-in-talks-with-turkey-and-iran-to-provide-food-water-official-idUSKBN18Y0Z8.

Reuters, Turkey and Russia to deploy in Syria's Idlib: Erdogan spokesman, *Reuters*, June 22, 2017, https://www.reuters.com/article/us-mideast-crisis-syria-turkey-russia/turkey-and-russia-to-deploy-in-syrias-idlib-erdogan-spokesman-idUSKBN19D1WI.

Review of preferential trade agreement between Iran and Turkey, rc.majlis.ir/fa/report/show/955255.

Sabah, Türkiye, Rusya ve İran'la ittifak arayışında olmalı, March 8, 2002, arsiv.sabah.com.tr/2002/03/08/p01.html.

Saeid Jafari, "Why Iran–Turkey Tensions Would Not Spiral Out of Control," *Al-Monitor*, March 13, 2017.

Saeid Jafari, Saudi-led Qatar blockade brings Iran, Turkey together, July 3, 2017, www.al-monitor.com/pulse/originals/2017/06/saudi-qatar-blockade-iran-turkey-convergence-crisis.html.

Seyyed Asadollah Athari, "Bazkhanie Ravaabete Do Keshvar pas az Jange Sard Iran ve Torkeye", *Rahbord*, No. 27 (Spring 2003).

Sharq Newspaper, Turkey, was the most beneficiary of Iran-Iraq war, *Shargh Newspaper*, September 22, 2012, old.sharghdaily.ir/pdf/91-06-29/vijeh/13.pdf.

Sigurd Neubauer, The curious timing of the Qatar crisis, *Al-Monitor*, June 6, 2017, www.al-monitor.com/pulse/originals/2017/06/timing-qatar-crisis-saudi-arabia-gcc-uae-egypt-diplomatic.html.

SSW, Between Turkey and Syria, Iran will Choose Assad, *Sobhe Sadegh Weekly*, July 18, 2011, www.asriran.com/fa/news/173230/ایران.سپاه-ارگان-میان-ترکیه-و-سوریه-اسد-را-انتخاب-می-کند.

Stephen Larabee and Alireza Nader, Turkish-Iranian Relations in changing Middle East (Santa Monica: Rand Corporation, 2011.

Suleyman Elik, Iran-Turkey Relations 1979–2011. Conceptualizing the Dynamics of Politics, https://jamestown.org/program/turkey-and-iran-expected-to-boost-security-cooperation/.

TNA, Erdogan Neo-Ottumanism or Democratization, *Tasnim News Agency*, April 18, 2017, https://www.tasnimnews.com/fa/news/1396/01/29/1382619/اردوغان-سازی-دموکراسی-یا-گری-نوعثمانی.

TNA, We Gave Rolling Pearl and Took Whey, January 6, 2015, https://www.tasnimnews.com/fa/news/1393/10/16/610481/درّ-غلتان-دادیم-و-آب-کشک-گرفتیم.

TT, "From Now On Turkey not to Allow Anti-Iran Activities on Its Soil-16-Point MoU Signed between Iran, Turkey", *Tehran Times*, December 10, 1998.

Turkey and Saudi Interference in Syrian Internal Affairs is a Big Mistake, *YJC*, February 12, 2016, www.yjc.ir/fa/news/5494918/ادخالتهای-عربستان-و-ترکیه-در-امور-سوریهاشتباه-بزرگی-است

Turkey can start new operation if necessary as Euphrates Shield ends: PM, *Hurriyet Daily*, March 29, 2017, www.hurriyetdailynews.com/turkeys-euphrates-shield-operation-in-syria-successfully-completed.aspx?pageID=238&nID=111401&NewsCatID=352.

Turkey is Worried About Syrian Kurds Autonomy, *Iranian Diplomacy*, June 10, 2016, www.irdiplomacy.ir/fa/page/1959934/ترکیه+نگران+از+خودمختاری+کردهای+سوریه.html.

Turkey, Iran, and Russia sign deal on safe zones in Syria, *The Nationale*, May 4, 2017, https://www.thenational.ae/world/turkey-iran-and-russia-sign-deal-on-safe-zones-in-syria-1.34497.

Turning Mosul into a Base Against Iran, *Paydari Melli*, December 6, 2015, paydarymelli.ir/fa/news/18822/تبدیل-موصل-به-پایگاهی-علیه-ایران

Unal Gundogan, Islamist Iran and Turkey 1979 1989 … 1979–1989: state pragmatism and ideological influences, *Middle East Review of International Affairs*, Vol. 7, No. 1 (March, 2003).

Yadullah Javani, Iran's Fundamental and Sincere Defending of Regional Nations, in Mohammad Khalilpour, A Different Story of Logic of Being in Syria (Tehran: Masir-e Aftab, 2017).

CHAPTER 9

Iran and Israel: Taking on the "Zionist Enemy"

Iran's policy toward Israel was, as Chap. 1 illustrated, guided by the Islamist idealism of Ayatollah Ruhollah Khomeini. The Supreme Leader declared Shiite Islam to be the new liberator of Jerusalem and the Holy Land from the so-called "Zionist enemy," or "Little Satan."

Theologians and Islamist scholars have engaged in a lively discourse about the alleged pending demise of the state of Israel. Khomeini's successor as Supreme Leader, Ayatollah Ali Khamenei, took it upon himself to develop a Quranic-based eschatology on the subject. He noted that the "Zionist regime" is built on weak pillars and thus is doomed to fail during the End of Days.[1]

In his view, this event would occur soon, when the Shiite messiah known as the "Twelfth Imam" or the "Mahdi" will appear to establish a global Islamic kingdom or caliphate. Khamenei supported his statement by quoting a verse from the Quran: "And we conveyed to the Children of Israel in the Scripture that, You will surely cause corruption on the earth twice, and you will surely reach [a degree of] great haughtiness. So, when the [time of] promise came for the first of them, we sent against you

[1] Khamenei: Zionist Regime is Doomed to Fail, *Khamenei.ir*, July 9, 2015, farsi.khamenei.ir/photo-album?id=24577.

F. Rezaei, *Iran's Foreign Policy After the Nuclear Agreement*, Middle East Today, https://doi.org/10.1007/978-3-319-76789-5_9

servants of Ours—those of great military might, and they probed [even] into the homes, and it was a promise fulfilled."[2]

Khamenei devoted considerable effort to the calculating and publicizing of the exact date of Israel's alleged demise. Based on verse 17: 5 in the Quran, he predicted a two-stage process of eliminating Israel, to be terminated 2022, equivalent to the year 5782 in the Hebrew calendar. By then all Jews would gather in Palestine as promised in the Quran 17: 104: "And We said after Pharaoh to the Children of Israel," Dwell in the land, and when there comes the promise of the Hereafter, We will bring you forth in [one] gathering [and we will punish everyone]."[3]

In a quote posted to Khamenei's official Twitter account on September 9, 2015, the Supreme Leader addressed Israel stating that "you will not see the next 25 years." He added that the "Jewish state would be hounded until it is destroyed. I'd say, firstly, you will not see next twenty-five years; God willing, there will be nothing as Zionist regime by next 25 years. Secondly, until then, struggling, heroic and jihadi morale will leave no moment of serenity for Zionists."[4]

Even pragmatic leaders such as Ayatollah Akbar Hashemi Rafsanjani, who was involved in procuring weapons from Israel during the Iran–Iraq war, occasionally used the Quranic prediction, apparently to drum up support for the nuclear program. During Friday Prayer meeting at Tehran University on December 14, 2001, he told the crowd: "If the Islamic world is also equipped with nuclear weapons, then the imperialists' strategy will reach a standstill because the use of even one nuclear bomb inside Israel will destroy everything." The international backlash was so enormous that the regime was forced to announce that Ayatollah Khamenei had imposed a fatwa on nuclear weapons.[5]

[2] Qur'anic Promise of Annihilation of Israel, https://www.tasnimnews.com/fa/news/1396/04/01/1443455/تحقق-زمان-درباره-تفسیری-نظریات-آخرالزمان-در-صهیونیسم-نابودی-قرآنی-وعده-اسرائیل-عذاب-بنی; Speaking for the 50 thousand Nationwide Basij Commanders, *Khamenei.ir*, farsi.khamenei.ir/speech-content?id=24552; Qoran, Surat Al-Isra (The Night Journey), 17 (4/5), https://quran.com/17.

[3] Qor'an, Surat Al-Isra (The Night Journey), 17:104, https://quran.com/17.

[4] Khamenei.ir, Khamenei's tweet on September 9, 2015, https://twitter.com/khamenei_ir/status/641520641724780544?ref_src=twsrc%5Etfw&ref_url=https:%2F%2Fwww.timesofisrael.com%2Fkhamenei-israel-wont-survive-next-25-years%2F.

[5] Akbar Hashemi Rafsanjani, Transcript of Rafsanjani Speech at Friday Prayer, Voice of the Islamic Republic of Iran Radio, December 14, 2001, http://www.globalsecurity.org/wmd/library/news/iran/2001/011214-text.html; IRNA, Bayanat dar Didar-e Zaerin-e Haram-e-Motahar Razavi [Speech in Mash'had for Imam Reza Pilgrims]. *Islamic Republic*

Unsurprisingly, the regime had consistently denied Israel's right to exist and harshly attacked Egypt and Jordan for making peace with Israel. Khomeini called the agreement between Israel and Egypt a conspiracy to legitimize Israeli aggression against Arabs and Palestinians. As the Supreme Leader put it, the "Camp David Agreement is a conspiracy to legitimize the Israeli aggression, which will consequently change the situation in favor of Israel and at the expense of the Arabs and Palestinians. Such a situation would not be acceptable to the people of the region."[6]

He also predicted that "progressive revolutions" would overthrow the "reactionary regimes," which protected the interests of colonialism and signed deals with Israel. To this end, he appealed to the Muslims: "O Muslims of the world and the oppressed people under the control of the oppressors, rise and unite against oppressors and defend Islam and your destiny. Do not be afraid of the great powers' rush because of this century, by the will of the God Almighty, is the century of defeating the arrogant powers and prevailing the right to the falsehood."[7]

Khomeini's dictum has been celebrated annually during the Quds Liberation Day, (Jerusalem Liberation Day) in Tehran and other cities. Organized by the IRGC, QF and the Basij, the display is known for its fevered parades and calls to liberate Quds with the "blood of the martyrs." In August 2012, Brigadier General Gholam Reza Jalali, the head of Iran's Passive Defense Organization, stated that "[Al-Quds Day] is a reflection of the fact that no other way exists apart from resolve and strength to eliminate the aggressive nature and to destroy Israel." Jalali added that the Muslim world was required to support the "oppressed people of Palestine" against "the Zionist usurpers" and that the Islamic Revolution was a "beacon of light" in this battle. On September 22, 2012, Major General Mohammad Ali Jafari, the commander of the IRGC predicted that a war

News Agency, March 22, http://news.iran-emrooz.de/more.php?id=11566_0_7_0_M; Khamenei, Bayanat dar Khotbe'haye Namaz Jom'e Tehran (Speech at Tehran Friday Prayers). *Khamenei.ir*, October 21, 2005.

[6] Ruhollah Khomeini, Sahifeh Noor, Vol. 4, p. 356: Khomeini.ir, Imam Khomeini and Camp David Plot, September 19, 2012, www.imam-khomeini.ir/fa/NewsPrint.aspx?ID=3062.

[7] Ruhollah Khomeini, Sahifeh Noor, Vol. 4, p. 356: Khomeini.ir, Imam Khomeini and Camp David Plot, September 19, 2012, www.imam-khomeini.ir/fa/NewsPrint.aspx?ID=3062; Fallah, Israel Nature from Imam Khomeini Perspective, August 17, 2012, www.al-falah.ir/portal/?pageID=910; Ruhollah Khomeini, Sahifeh Noor, Vol. 14, p. 63; Rouhollah.ir, Sahifeh Imam Khomeini, Vol. 15, pp. 168–172.

with Israel would soon break out. In his view, Iran would eradicate Israel, which he referred to as a "cancerous tumor."[8]

Members of the Haqqani School were also keen on seeing the prophecy of Israel's destruction come through, a theme which preoccupied their spiritual leader Ayatollah Mohammed-Taqi Mesbah Yazdi. Many in the Haqqani network, including President Mahmoud Ahmadinejad, used the prediction in public speeches. In December 2005, in a speech to commemorate the nineteenth anniversary of Khomeini's death, Ahmadinejad told the crowd that "you should know that the criminal Zionist regime will soon disappear off the geographical scene." By grabbing more world headlines, the president described Israel as "a cancerous tumor" that had to be wiped off the map; in another version, he referred to Israel as a "stain on the map."[9]

Ahmadinejad proudly declared that the world could look forward to a time when Israel would be erased from the pages of history: "You should know that this slogan, this goal, can certainly be achieved." Sounding apocalyptic, Ahmadinejad expressed the belief that the fate of Israel would be determined in a broader war that Islam was destined to wage against the infidels, a theme unveiled in many of his addresses.[10]

[8] Shafaghna, War With Israel Would Soon Break out, *Shia News Association*, September 22, 2012, www.shafaqna.com/persian/component/k2/item/22997 فرمانده-سپاه-دشمني-رژيم-صهيونيستي-با-ايران-به-درگيري-فيزيكي-ختم-ميشود; ManotoTV, IRGC Commander Called for Readiness for War with Israel, September 22, 2012, https://www.youtube.com/watch?v=GFtYnk6a3ys.

[9] Ali M. Ansari, Iran Under Ahmadinejad: The Politics of Confrontation (London: Routledge, 2007); Ethan Bronner, Just How Far Did They Go, Those Words Against Israel? *The New York Times*, June 11, 2006; Nazila Fathi, Wipe Israel 'Off the Map' Iranian Says. *The New York Times*, October 27, 2005; IRNA, Ahmadinejad: Those Who Recognize Israel Will be Punished by the Wrath of the Islamic Nations. *Islamic Republic News Agency*, October 26, 2005, http://www.president.ir/fa/2288; Ewen MacAskill and Chris McGreal, Israel Should Be Wiped Off Map, Says Iran's President. *The Guardian*, October 27, 2005; Reuters, 2008. Ahmadinejad Says Israel Will 'Disappear'. *Reuters*, June 3, http://www.reuters.com/article/2008/06/03/us-iran-israel-usa-idUSL0261250620080603; Amir Taheri, The Persian Night: Iran Under the Khomeinist Revolution (New York: Encounter Books, 2010).

[10] Fars-News-Agency, Ahmadinejad Remarks on Zionist Regime Provoked Wrath of Zionism Supporters. *Fars News Agency*, October 27, 2005, http://www.farsnews.com/newstext.php?nn=8408040372; IRDC, Darbareye Roze Jahaniye Qods [About International Day of Qods]. *Islamic Revolution Documents Center*, July 23, 2014, http://www.irdc.ir/fa/content/47167/print.aspx; Amir Taheri, Sympathy from the Devil. *New York Post*, September 23, 2007.

Gholam Hossein Elham, another member of Haqqani School, declared that "Israel is condemned to destruction, and the world without Israel is a world that its signs can be seen."[11]

At the operation level, the Revolutionary Guards, which, as noted in Chap. 1, took on the task of revolutionary expert, were also charged with Khamenei's edict of "hounding Israel" until it was destroyed. Created in 1982, the Lebanese Hezbollah was a perfect tool for both these missions. As noted in Chap. 1, the IRGC-QF developed the strategy of low-level asymmetrical war by using proxies. Having learned from the Arab failure to destroy Israel in a series of major wars, the Guards opted for destruction by attrition; that is, "through persistent warfare that gradually weakens the Jewish state" without risking a direct clash with the United States or Israel. Significantly, the group was tasked with muqawama, "resistance" to the Jewish states, and its eventual annihilation.[12]

Lebanese theologians added a religious dimension to the muqawama by borrowing from Ayatollah Khamenei, who was highly revered by the Shiite community. Hashem Saffiedian, the head of the Executive Council of Hezbollah, credited Khamenei with providing general spiritual guidance. Hujjat al-Islam Ali al Khourani, a Lebanese scholar, propagated the theory that the 25 years would culminate in a final battle in Lebanon.[13]

Using the Lebanese Hezbollah to target Israel along its northern border was the method of choice. IRGC-QF supplied most of Hezbollah's arms, including modern antitank weapons and its thousands of rockets. Its advisors spent years helping the militants train and build fortified positions throughout southern Lebanon, and Hezbollah fighters attacked Israel Defense Force (IDF) soldiers stationed in the so-called separation zone in the south. The IRGC-QF considered it a triumph when the then Prime Minister Ehud Barak ordered the IDF out of Lebanon in 2000. After Israel's departure, the Guards decided to turn Hezbollah into a semi-professional military force equipped with heavy weapons and missiles while retaining its unique civilian characteristics.[14]

[11] RFI, The Second Round of Rejection of the Killing of the Jews, January 27, 2009, www1.rfi.fr/actufa/articles/109/article_5152.asp.

[12] Nadav Pollak, The Transformation of Hezbollah by its Involvement in Syria (Washington, DC: Washington Institute for Near East Policy, 2015), 5.

[13] Shiite News, "Islamic Resistance," *Shiite News*, September 14, 2015; Shimon Shapira "Hezbollah's Veneration of Iranian Leader Ali Khamenei." *Jerusalem Center Blog*, June 20, 2011; Amir Taheri, Hezbollah and Iran? Who Pays the Price," *Asharq al Awsat*, June 10, 2011.

[14] Laboratory of Asymmetry: The 2006 Lebanon War and the Evolution of Iranian Ground Tactics Lindemann, Marc, Military Review May–June 2010.

This type of hybrid warfare proved difficult for the IDF, not least because Hezbollah dispersed its fighters and armaments among the population. The guerrilla-style operations proved to be another effective way to harass Israel. Qassem Suleimani devised a plan for cross-border actions. In the six years since 2000, Hezbollah executed some 200 cross-border attacks, including artillery fire, raids, and attempts at kidnapping. On July 12, 2006, Hezbollah launched an assault resulting in the killing of three Israeli soldiers and the abdication of two more. Hezbollah's Secretary General Hassan Nasrallah considered the attack to be routine, but Israel responded with force. The 33-day war killed some 1,300 Lebanese and displaced one million, and severely damaged the infrastructure. Israel's casualty rate stood at 165, but the IDF was surprised by the barrage of short and medium-range rockets and missiles that reached beyond Haifa and displaced up to 500,000 civilians.

Ali Akbar Mohtashami-Pur, by then the Secretary-General of the "Intifada Conference," an organization created to disrupt the Oslo peace process, told the Shargh newspaper that the Israelis suffered because of Hezbollah's arsenal of rockets and missiles, including the Zelzal-2 missile. According to specifications the missile had a range of 200 km range, a length of 8.46 m, a diameter of 0.61 m, and was able of carrying a 600-kg warhead.[15]

He also disclosed that the Guards transferred this and other weapons in response to Hezbollah's request "to support its operation against Israel." Hezbollah also received Russian, Chinese and Iranian designed air-to-surface missiles, including the Strela 2, Man-Portable Air Defense Missile System (MANPAD). Unconfirmed reports at the time spoke of several IRGC-QF personnel killed in the fighting.[16]

[15] Sharq Newspaper, "Interview with Ali Akbar Mohtashami." *Shargh Newspaper*, 2006-08-03, ww25.sharghnewspaper.com/850512/html/polit.htm#s453504; IRARTESH, Zelzal Rocket the most accurate Iranian rocket, February 19, 2013, irartesh.ir/post/581/زلزال_دقیق_ترین_راکت_ایرانی.html.

[16] Katz, Yaakov (August 4, 2006). "Iran: We supplied Zelzal-2 to Hizbullah". *Jerusalem Post*. Archived from the original on July 6, 2013. Retrieved 2006-08-13; "Israel's war With Iran" by Ze'ev Schiff, *Foreign Affairs*, November–December 2006, p. 23; JDW, "Iran answers Hizbullah call for SAM systems." *Jane's Defense Weekly*, 2006-08-07; "War Dead Flown to Iran". *The New York Sun*, July 24, 2006. Archived from the original on January 4, 2011; Ynet, Report: Iranian soldiers join Hizbullah in fighting, *Ynet News*, July 24, 2006, www.ynetnews.com/Ext/Comp/ArticleLayout/CdaArticlePrintPreview/1,2506,L-3280446,00.html; AW, "Iranian Advisers Influence Course of Lebanon/Israel Conflict". *Aviation Week*, August 14, 2006. Archived from the original on 29 April 2011.

Both the Iranians and Hezbollah were quick to claim victory in the war. The Supreme Leader that Iran supported the resistance movement against the Zionist regime in the 33 days' war. Hassan Nasrallah, Hezbollah's secretary-general, said: "I explicitly say that since 1982, Iran has provided all material, spiritual and political support. The resistance movement that resisted and won the 2006 Tamoz War [33 days of 2006] could not resist and win if it did not receive support from Iran."[17]

Rouhani's Quest to Change the Attitude to Israel: Realism and Idealism on a Collision Course

As part of the normalization plan, President Hassan Rouhani tried to balance the theological dictates of the Islamic Republic with the realpolitik needs of his policy. Rouhani and Foreign Minister Mohammad Javad Zarif understood that the extremely hostile anti-Israel rhetoric besmirched Iran's international reputation and jeopardizing the normalization project. This hostility was also complicating the nuclear negations because opponents in the United States tried to derail the deal by emphasizing the "genocidal threats" of Iranian leaders. Rouhani's first order of business was to tone down the commentary on the Quds Day. During a TV interview, he stated that "Quds Day, which is in memorial of Imam [Khomeini], is a day that people present the unity of Islam against any oppression or aggression." Notably, he mentioned the Land of Palestine once and made no references to the Israel, or the "destruction of the Zionist enemy."[18]

Rouhani's government faced an equal challenge in taking on the Holocaust denial tradition, which President Ahmadinejad first embarked on soon after coming to power. On December 14, 2005, during a tour of the Zahedan province, Ahmadinejad articulated the view that the Holocaust was a myth created to justify the formation of Israel—described as "a move by the oppressor of the world [the United States] against the Muslim world." Holocaust denial, a long-standing tradition among fringe scholars and anti-Semitic politicians in the West, became Ahmadinejad's tool of choice. In his view, "They have fabricated a legend, under the name of the Massacre

[17] Ammariyon, The Victory of Lebanon's Resistance in 33-day war was Obtained by Iran's Support, *Ammariyon.ir*, http://www.ammariyon.ir/fa/news/18178/پیروزی-مقاومت-لبنان-در-جنگ-33-روزه-با-حمایت-ایران-حاصل-شد.

[18] Aparat, Interview with Hassan Rouhani in Qods Day, *Aparat*, www.aparat.com/v/nCNby/مصاحبه_حسن_روحانی_در_روز_قدس_سال_96.

of the Jews, and they hold it higher than God himself, religion itself and the prophets themselves…If somebody in their country questions God, nobody says anything, but if somebody denies the myth of the massacre of Jews, the Zionist loudspeakers, and the governments in the pay of Zionism will start to scream."[19]

To prove that the Holocaust did not occur, Ahmadinejad organized conferences which featured assorted Holocaust deniers. After Ahmadinejad left office, the IRGC's cultural organs took over the Holocaust denial franchise. Zarif waited no time to denounce Ahmadinejad's penchant for Holocaust denial. Without mentioning the name of the former president, in September 2013, he wrote on his Twitter account, "Iran never denied it. The man who was perceived to be denying it is now gone." To assure the world that the new government was not anti-Israel he ended his tweet with a "Happy New Year."[20] On May 6, 2014, Zarif was summoned to Majlis and questioned about his foreign policy opinions in a session that was broadcasted live on state radio. The foreign minister defended his stand on Holocaust denial: "But my colleagues and I are telling the world Iran is opposed to anti-Semitism and genocide."[21]

Despite such declarations, the Rouhani government could not prevail upon the Guards which, as noted above, had a virtual monopoly on dealing with Israel. The competition for Holocaust cartoons, the brainchild of Masoud Shojaei-Tabatabaii, the head of House of Cartoons, created a serious showdown. After the French satirical magazine Charlie Hebdo published a cartoon depicting the Prophet Mohammed, Shojaei-Tabatabai commissioned the Owj Media & Art Institute, and the Sarcheshmeh Cultural Complex, to launch an international competition for Holocaust denial cartoons. Owj Media was owned by the Revolutionary Guards and Islamic Republic Iran Broadcast (IRIB), which is controlled by the Guards, and the Sarcheshmeh Cultural Complex received funds from the Guards. The two centers offered

[19] Quoted in Simon Mabon, Saudi Arabia and Iran: Power and Rivalry in the Middle East (I.B. Tauris, 2016).

[20] MNA, Zarif; We got comfort from Netanyahu, *Mehr News Agency*, May 6, 2014, www.mehrnews.com/news/2286013/ظریف-آسایش-را-از-نتانیاهو-گرفته-ایم-کریمی-قدوسی-دلواپسی-رمز; Zarif, personal Twitter Account, September 5, 2013, https://twitter.com/jzarif/status/375617854214660097?lang=en.

[21] MNA, Zarif; We got comfort from Netanyahu, *Mehr News Agency*, May 6, 2014, www.mehrnews.com/news/2286013/ظریف-آسایش-را-از-نتانیاهو-گرفته-ایم-کریمی-قدوسی-دلواپسی-رمز; Zarif, personal Twitter Account, September 5, 2013, https://twitter.com/jzarif/status/375617854214660097?lang=en.

cash prizes for the winners. Over 150 cartoons made up a May 2016 exhibition at the Islamic Propaganda Organization in Tehran.[22]

Pronouncing the competition to be a great success, Shojaei-Tabatabaii, launched another international context. International artists sent cartoons that mocked the genocide and the suffering of the Jews were put on exhibit by the Arts Bureau in Tehran. Over 150 artists from 50 countries were taking part in the Third International Holocaust Cartoons Competition on the premises of Iran's Arts Bureau in Tehran. Shojaei-Tabatabai added a section that depicted the Israeli Prime Minister Benjamin Netanyahu and Adolf Hitler have been highlighted in the caricature section.[23]

In an effort of damage control, Foreign Minister Zarif attempted to distance the administration of President Hassan Rouhani from the contest. Speaking with the New Yorker on April 25, 2016, Zarif said the event was sponsored by private, nongovernmental organizations. As Zarif put it, "It's not Iran. It's an NGO that is not controlled by the Iranian government. Nor is it endorsed by the Iranian government."[24]

For the traditionalists, however, the rhetorical vilification of Israel was not only a theological mandate but a platform for scoring points against Rouhani whom they accused of being "soft" on the Zionists. As typical of the negotiated political order, Israel bashing became a sign of defiance of the Rouhani normalization project. For instance, during the Artesh Day on April 19, 2017, the Guards displayed surface-to-surface ballistic missiles. Among them was the Zolfaghar and the Ghadr with a range of 2,000 km; the latter could reach the American bases in the region and Israel.[25]

[22] Iran Caricator, The opening ceremony of the 2nd International Cartoon Contest and the Holocaust Cartoon, May 14, 2016, art.irancartoon.ir/content/مراسم-افتتاح-2مین-مسابقه-بین-المللی-کارتون-و-کاریکاتور-هولوکاست-برگزار-شد/.

Iran Caricature, The Opening Ceremony of the 2nd Holocaust International Cartoon Contest, *Iran Caricature*, May 14, 2016, art.irancartoon.ir/content/مراسم-افتتاح-2مین-مسابقه-بین-المللی-کارتون-و-کاریکاتور-هولوکاست-برگزار-شد/.

[23] The organizers increased the winning awards from $20,000 to $50,000 Holocaust Denial: Iran Holocaust Cartoon Exhibition, https://www.ushmm.org/fa/iran-cartoon-fa; IW, The Holocaust Cartoons and Zarif's Lies, *Iran Wire*, https://iranwire.com/en/features/1781.

[24] Robin Wright, Iran's Javad Zarif on the Fraying Nuclear Deal, U.S. Relations, and Holocaust Cartoons, *The New Yorker*, April 25, 2016, www.newyorker.com/news/news-desk/irans-javad-zarif-on-the-fraying-nuclear-deal-u-s-relations-and-holocaust-cartoons.

[25] Times of Israel Staff, "Unveiling clock showing 8411 days left for Israel, Iranians rage against Jewish state," *Times of Israel*, June 24, 2017, 3.

The Guards made the point clear; the missiles were painted with a slogan "death to Israel." Rouhani wrote a letter to Ayatollah Khamenei demanding the removal of the slogans, but the Supreme Leader who used the same line in his speeches was not forthcoming.[26] Thus enabled, the IRGC paraded missiles with the same slogan on Quds Day on June 23, 2017.

Reaching a new level of animus towards Israel, the rally also featured a huge digital "Doomsday Clock" indicating that Israel would cease to exist in 8441 days, that is 25 years. The participants evoked the Supreme Leader's prophecy, changed "Death to Israel" and proclaimed that the destruction of the Jewish state was "the top priority of the Muslim world."[27]

Faced with such an outpouring of anti-Israel sentiments, President Rouhani responded by criticizing the Guards for the provocative message on the missiles. In a debate before the May 19 election, he accused the hardliners of trying to sabotage the nuclear deal. "We saw how they wrote slogans on missiles and showed underground (missile) cities to disrupt the JCPOA." The president added that the IRGC and other hardliner opponents were divisive and harmed the chances of the society to benefit from the nuclear deal.[28]

He ended by stating: "We succeeded wherever we were united; we were together; from there we were able to use the benefits of the nuclear agreement.[29]

[26] Rahkar News, Rouhani Called for Removing "Death to Israel" Writing on the Missiles, March 26, 2016, www.rahkarnews.ir/fa/doc/news/20794/ روحانی-خواستار-حذف-شعار-هک-شده-مرگ-اسرائیل-روی-موشک-های-بالستیک; Jame'e Khabar, Iran Called "Death to Israel", April 18, 2017, https://www.jamehnews.com/8419/-روزنامه-انگلیسی-ایران-در-روز-ارتش-شعار-مرگ-بر-اسرائیل-سر-داد1

[27] Times of Israel Staff, "Unveiling clock showing 8411 days left for Israel, Iranians rage against Jewish state," *Times of Israel*, June 24, 2017, 3; Rahkar News, Rouhani Called for Removing "Death to Israel" Writing on the Missiles, March 26, 2016, www.rahkarnews.ir/fa/doc/news/20794/روحانی-خواستار-حذف-شعار-هک-شده-مرگ-اسرائیل-روی-موشک-های-بالستیک; Jame'e Khabar, Iran Called "Death to Israel", April 18, 2017, https://www.jamehnews.com/8419/روزنامه-انگلیسی-ایران-در-روز-ارتش-شعار-مرگ-بر-اسرائیل-سر-داد-1.

[28] TNA, The End of Four-Percent Government, *Tasnim News Agency*, May 6, 2017, https://www.tasnimnews.com/fa/news/1396/02/16/1399228/پایان-دولت4--درصدی-ها

[29] Tnews, Rouhani Speech Gives excuse to the West, tnews.ir/news/fab485219452.html.

Bolstering the Anti-Israel Proxies: Hezbollah, Hamas, and Palestinian Islamic Jihad

Following the 2006 war, the Revolutionary Guards replenished Hezbollah's arsenal and added new strategies to enhance its capability. Fighting in Syria improved Hezbollah's military experience. According to Israeli military sources, Hezbollah gained valuable knowledge by watching the Russians and the Iranians. Hezbollah commanders learned how to plan and execute complex operations. They also acquired advanced knowledge in Russia's intelligence and electronic warfare (IEW) capabilities when the latter deployed the TU-214R, intelligence, surveillance, and reconnaissance (ISR) aircraft. Hezbollah operatives were also able to study the land-based Zoopark-1 radar system and the RB-341 V (Leer-3), an electronic warfare system equipped with an unmanned air vehicle (UAV)—Orlan 10. Observing the Russians and Iranians drones, the Hezbollah improved its skills of managing its own fleet of drones, estimated at of 200 Iranian-made UAVs (Unmanned Aerial Vehicle) capable of low-altitude flights to evade detection by radar. The Hezbollah drones could photograph sites or attack strategic civilian and military bases in Israel.[30]

On October 14, 2012, Hezbollah sent an Iranian-made drone dubbed "Ayoub" into Israeli airspace to photograph secret military bases inside the occupied territories. According to a report by *The Times*, "The drone, which was airborne for three hours before being intercepted by an F-16 jet, is believed to have transmitted pictures of preparations for Israel's biggest joint military exercise with the US army, which began last week, as well as ballistic missile sites, main airfields and, possibly, its nuclear reactor in Dimona."[31]

General Ahmad Vahidi, Iran's Minister of Defense and former chief of the QF, acknowledged that the drone used Iranian technology. "Hezbollah's launch of a drone into Israeli airspace last week proves the Islamic republic's military capabilities. Great job by Hezbollah. The era in which the Zionist regime could think it has regional supremacy is over." Vahidi

[30] Nadav Pollak, The Transformation of Hezbollah by its Involvement in Syria (Washington, DC: Washington Institute for Near East Policy, 2015), 5; Roi Kais, Hezbollah has fleet of 200 Iranian-made UAVs, November 25, 2013, www.ynetnews.com/articles/0,7340,L-4457653,00.html.

[31] Uzi Mahnaimi, Security shambles as Hezbollah drone spies on Israeli army, *The Times*, October 14, 2012, https://www.thetimes.co.uk/article/security-shambles-as-hezbollah-drone-spies-on-israeli-army-n6kxp77dnz7.

announced that Iran currently had drones with much more advanced technology than the drone that has recently been flown by Hezbollah over Israel. "Of course, the UAV technology recently flown by the Lebanese Resistance Force over the regime occupied Quds and amazed the enemy was not the latest Iranian technology." Vahidi added that "The Hezbollah's exponential power by flying the Ayoub UAV's drones over the sky of the Zionist regime destroyed the grandeur of this fake regime and hit the Israelis hard. The Zionist regime had propagated its Iron Dome air defense system, but this drone disgraced them and took their dignity because they carry out raids on their enemies through psychological operations and emboldening their military power."[32]

In another benefit of the war, Hezbollah has expanded its already significant military infrastructure in Syria, including weapon storage, training facilities, and logistical bases. Since 2011, Hezbollah has built bases in territories that it had helped to liberate from anti-Assad rebels. For instance, after its victory in al-Qusayr in June 2013, Hezbollah reportedly built a base near the city, close to the Syrian-Lebanese border. Military experts assumed that the base could permanently house some of its 3,000-odd fighters in Syria. It was also planned to serve a training facility and a coordination point with IRGC operatives.[33]

Acquiring new military capabilities, however, came with a cost. Out of the estimated 20,000 Hezbollah fighters (and 25,000 reservists), around 1,200 were killed and 5,000 wounded, more than the tally from the militia's 1985–2000 war with Israel in southern Lebanon, in which 1,248 Hezbollah fighters were killed and at least a thousand wounded.[34]

The mounting losses created dissent among the families of the casualties, the rank and file, and even some senior commanders. Indeed, Mustafa Badreddine, the Hezbollah commander in Syria, was said to have opposed the open-ended use of Hezbollah in Syria and the use of the "Arab Shiites" as cannon fodder. Badreddine died on May 13, 2016, under disputed

[32] Entekhab, Iran Declared: Hezbullah Drone was Iranian, *Entekhab.ir*, www.entekhab.ir/fa/news/80118/ایران-تایید-کرد-هواپیمای-بدون-سرنشین-حزب-الله-ایرانی-بوده-است

[33] Stratfor, "Hezbollah's Shot at Permanency in Syria," *Stratfor*, April 6, 2016, https://www.stratfor.com/analysis/hezbollahs-shot-permanency-syriaک BI, Satellite imagery reveals Hezbollah's play for permanent bases in Syria, *Business Insider*, April 6, 2016, www.businessinsider.com/satellite-imagery-reveals-hezbollahs-play-for-permanent-bases-in-syria-2016-4.

[34] Dan De Luce, Syrian War Takes Rising Toll on Hezbollah, *Foreign Policy*, July 9, 2015, foreignpolicy.com/2015/07/09/syrian-war-takes-rising-toll-on-hezbollah/.

circumstances. While the official version held that he was killed in a rebel attack near a base in Damascus International Airport, Arab and Israeli intelligence services concluded that Hassan Nasrallah, acting under pressure from Qassem Soleimani, ordered the hit.[35]

For Nasrallah, however, the casualties were a fair price to pay for following the IRGC-QF line. Israeli intelligence concluded that Hezbollah's military strength surpassed the pre-2006 level. Hezbollah's weapons arsenal grew from 33,000 rockets and missiles before the 2006 war to an estimated 150,000. It has also diversified its firing capabilities: "underground launchers, launchers hidden in nature reserves and portable launchers aboard trucks and commercial vehicles." The IDF estimates that in the next war, Hezbollah would be able to field a hybrid force boasting some 30,000–33,000 solders, both standing, and reserve. The group could fire up to 1,500 rockets and missiles a day, compared to 200 in the 2006 war. Israeli sources listed Grad rockets with a range of 40 km, Fajr rockets with a range of 75 km, Zelzal missiles with a range of 200 km, Fateh, and M-110 missiles with a range of 250 km, and Syrian D-model Scud missiles with a range of 700 km. Additionally, Hezbollah possessed Chinese-made C-802 missiles and Yakhont missiles with a longer range, better precision, and diverse launching options.[36]

Hezbollah has come to possess an array of unmanned aerial vehicles, as noted above. These include drones that can carry explosives, and suicide drones. During the fighting in Syria, the militants used drones with attack capabilities. Hezbollah's antitank-missile stock had increased significantly. In addition to the Sagger missiles, rocket-propelled grenades, Fagot missiles and MILAN missiles, it boasted advanced anti-aircraft missiles and advanced Kornet missiles that can penetrate the Merkava 4 tank.[37]

Guards' engineers helped Hezbollah to set up underground factories to manufacture rockets. Imam Hossein University in Tehran, the main Revolution Guards university, has trained dozens of Hezbollah students in engineering and rocket technology. Sources disclosed that the IRGC helped to develop Hezbollah's indigenous manufacturing capabilities after

[35] Judah A. Gross, IDF chief: Hezbollah commander killed by his own men last year, *Times of Israel*, March 21, 2017, www.timesofisrael.com/idf-chief-hezbollah-commander-killed-by-his-own-men-last-year/.

[36] Amos Harel and Gili Cohen, Hezbollah From terror group to army, *Haaretz*, June 12, 2016, www.haaretz.com/st/c/prod/eng/2016/07/lebanon2/.

[37] Ibid.

the Israelis bombed its rocket making factories and in Sudan and the supply routes via Syria.[38]

Despite the UN Resolution 1701, which forbids Hezbollah to maintain any military presence south of the Litani river, the build has continued relentlessly. Israel has criticized the United Nations Interim Force in Lebanon (UNIFIL) charged with overseeing the implementation of the resolution, but not much has changed. Legal experts noted that hybrids like Hezbollah have an upper hand over a conventional army bound by the Geneva Conventions. The latter "struggle with the virtually impossible mission of avoiding civilian casualties while fighting the deeply embedded militants."[39]

In the IRGC-QF playbook, a revamped Hezbollah was to serve as a major deterrence to a potential Israeli attack. Iranians believed that Hezbollah's rockets would inflict serious physical damage on Israel, terrorize Israeli cities, and draw Israeli ground troops into a tough confrontation. Western scholars have agreed with this view. As one of them noted, by learning the lessons of 2006, Hezbollah perfected both "deterrence by punishment" and "deterrence by denial." The former counted on the expanded array of rockets and missiles to exact the highest possible cost in a future conflict. Hezbollah bragged that its new arsenal could hit Tel Aviv and beyond. The increased precision was a clear indication that the militant group aimed at Israel's Achilles heel, its civilian population. As for the latter, they touted the intricate network of tunnels and hideouts, including military posts designed to hinder a conventional army while hiding among civilians.[40]

[38] TNA, Reviewing how the 'Israeli Mercado Tank was Destroyed' at the University of Revolutionary Guards, https://www.tasnimnews.com/fa/news/1396/05/01/1467751/ بررسی-نحوه-انهدام-تانک-مرکاوای-اسرائیل-در-دانشگاه-افسری-سپاه-پاسداران.

[39] John R, Davis, Jr., "Defeating Future Hybrid Threats: The Greatest Challenge to the Army Profession of 2020 and Beyond," *Military Review*, September–October, 2013; Amos Harel, "Israel Launches Preemptive Strike in Psychological War with Hezbollah," *Haaretz*, June 24, 2017.

Gilli Cohen, "Israeli Army Intelligence Chief: Hezbollah Is Setting Up a Weapons Industry in Lebanon With Iranian Know-how," *Haaretz*, June 23, 2017.

[40] Daniel Sobelman, "Learning to Deter: Deference Failure and Success in the Israeli-Hezbollah Conflict, 2006–2016," Learning to Deter: Deterrence Failure and Success in the Israeli-Hezbollah Conflict." *International Security*, 41, issue 3, 2017; Ivan Arreguín-Toft, How the Weak Win Wars: A Theory of Asymmetric Conflict (Cambridge: Cambridge University Press, 2005); Ivan Arreguín-Toft, "Unconventional Deterrence: How the Weak Deter the Strong," in Paul, Morgan, and Wirtz, Complex Deterrence, pp. 213–214.

If Hezbollah turned out to be the perfect proxy, IRGC-QF relations with Hamas were complicated by sectarian differences and the pull of rival regional alliances. In the 1980s, Iran has provided 10 percent of Hamas's funding; this amount increased annually until it reached $30 million in 1990s. In the early 2000s, Iran provided Hamas with $22 million, over a quarter of the Hamas budget. To evade the government's scrutiny, the IRGC-QF set up a complex structure of relations with both the political wing of Hamas which had ruled the Gaza Strip since 2006 and its military wing, the Izzadin al Qassam Brigades. Hezbollah, which Hamas viewed as an operational model, played an important role in this process, by training Hamas militants in Lebanon or helping with arms smuggling. But Hamas never matched the degree of fealty to Iran as the much smaller Palestinians Islamic Jihad (PIJ), a virtual QF creation.[41]

Iran has also supplied military assistance to Hamas in Gaza, including technology needed to build Fajr-5 rockets with a range of 47 miles and carry a warhead of nearly 400 pounds which could target Tel Aviv. Ali Jafari, the Guards chief, was quoted as saying "Gaza is under siege, so we cannot help them. The Fajr-5 missiles have not been shipped from Iran. Its technology has been transferred and (the missiles are) being produced quickly."[42]

Due to Iran's help, by the late 2000s, Hamas became a formidable foe of Israel. During the three Israeli operations in Gaza—the 2008–2009 Operation Cast Lead, the 2012 Operation Pillar of Defense, and the 2014 Operation Protective Edge—Izzadin al Qassam made good use of the Iranian-provided hardware. In fact, all three operations were precipitated by rockets fired from the Gaza Strip. During the 2012 conflict, the militants announced firing the Fajr-5 at Tel Aviv. On the ground, the IDF was hampered by an extensive network of tunnels located in the heavily populated Strip. Like Hezbollah, Hamas and PIJ located key military facilities under public buildings, hospitals, and schools.[43]

[41] DW, Iran Denied Sending Fajr-5 Missile to Hamas, *DW*, November 18, 2012, www.dw.com/fa-ir/ایران-ارسال-موشک-فجر۵--برای-حماس-را-تکذیب-کرد/a-16387880.

[42] FNA, Commander-in-Chief of the Guards: Iran did not send weapons to Gaza, but Technology Transfer, *Fars News Agency*, www.khabaronline.ir/detail/259262/Politics/military.

[43] FNA, Commander-in-Chief of the Guards: Iran did not send weapons to Gaza, but Technology Transferred, *Fars News Agency*, www.khabaronline.ir/detail/259262/Politics/military.

The civil war in Syria strained the relation between Iran and its Hamas proxy. While Hezbollah followed the Revolutionary Guards into supporting Assad, some Hamas combatants found their way into anti-Assad militias. At least on one occasion, in the town of Quasar, the Izzadin al Qassam militants fought the IRGC-QF-Hezbollah contingent which assisted the Syrian army. In 2012, Khaled Mashal, the head of the political bureau of Hamas, refused to support Iran's policy in Syria and left Damascus, his longtime location. The rift deepened when Mashal, by then residing in Qatar, took the side of the Yemeni President Abdrabbuh Mansour Hadi against the Houthis. There were financial repercussions to this show of independence. In 2012, Iran dropped its financial assistance to Hamas and reduced the funding to the military wing from $150 million in 2012 to $60 million in 2013. The IRGC also imposed tougher contains on the aid, including a request for video proof of attacks on Israeli targets.[44]

Pivoting back to Iran became more urgent after Protective Edge which left Hamas under extreme financial pressure. Saudi Arabia, aligned with Egypt, declared the Moslem Brotherhood, of which Hamas is a branch, to be beyond the pale and cut the purse strings. Both the political and military leadership agreed to the pivot. In the words of a senior Hamas official "The Arab world is undergoing a redrawing of its political map. The Syrian regime is achieving success on the ground against its opponents, Saudi Arabia is preoccupied with its war in Yemen and Turkey is restoring relations with Israel. As a result, Hamas fears a continuation of its isolation due to regional developments. It is thus consciously resuming its rapprochement with Iran out of its leadership's full conviction that doing so was the best option to safeguard its strength militarily and politically." He added that the military wing of Hamas "strongly endorsed the resumption of said relations, especially in light of the urgent need for funds and military support that only Iran can provide." In June 2016, the veteran Hamas operative Mousa Abu Marzouk explained that the "the support offered by Iran to the Palestinian resistance—be it in logistics, training or funds—is unmatched and beyond the capabilities of other countries. Iran's support and backing to the resistance and the Palestinian cause are clear,

[44] Richard Davis, Hamas, Popular Support and War in the Middle East: Insurgency in the Holy Land (New York: Routledge, 2016); Matthew Levitt, Hamas: Politics, Charity, and Terrorism in the Service of Jihad (Yale University Press, 2006), pp. 143–144; Jodi Vittori, Terrorist Financing and Resourcing (Palgrave Macmillan, 2011), pp. 72–74, p. 193.

explicit and equal to the stance of Arab, Muslim and free peoples of the world who back and support the resistance."[45]

Financially, the Iranian pivot paid off quite handsomely. According to Herzel Halevi, the head of Israeli military intelligence, Iran paid $50 million to Hamas and $70 million to Palestinian Islamic Jihad (PIJ).[46]

Lavishing cash on the Lebanese and Palestinian proxies to shoulder the burden of resistance against Israel was, as noted in Chap. 1, a rational decision in the service of the Khomeini-mandated revolutionary export. But the Israeli military developed a new deterrence strategy based on advanced technology, which the Revolutionary Guards could not have anticipated, questioned the future of this substantial investment.

Cumulative Deterrence and "War Between Wars": IDF's Response to Asymmetrical Warfare

After decades of experience with asymmetrical warfare, the IDF adjusted the conventional strategy of deterrence. Major General Doron Almog explained that classic Cold War deterrence was of little use in fighting Hezbollah, Hamas, and the PIJ. Instead, the military opted for the so-called "cumulative deterrence" built on "victories achieved over the short, medium, and long terms that gradually wear down the enemy." The IDF believed that wearing the enemy down would require "multilayered, highly orchestrated effort to inflict the greatest damage possible on the terrorists and their weapon systems, infrastructure, support networks, financial flows, and other means of support. It would demand excellent intelligence, a broad coalition, and a globalized network that would facilitate the exchange of vital information and encourage transparency." Above all, "it would require cutting-edge technology."[47]

Vital for the success of cumulative deterrence required the strategy of "war between the wars" described by senior military officials as "small signature raids" such as bombings, raids, and other clandestine activities. By its admission, the IDF has carried out hundreds of such acts against

[45] Hazem Balousha "Why Hamas resumed ties with Iran," *Al-Monitor*, June 29, 2016; Hanin Ghaddar, The Marriage and Divorce of Hamas and Hezbollah, *Wilson Center*, https://www.wilsoncenter.org/article/the-marriage-and-divorce-hamas-and-hezbollah.

[46] Yonah Jeromy Bob, "Massive Iranian Funding for Anti-Israel Terror Group Revealed," *Jerusalem Post*, June 22, 2017.

[47] Doron Almog, "Cumulative Deterrence and the War on Terrorism," *Parameters*, Winter, 2004–2005, 4–19.

Hezbollah, Hamas, and targets in Syria. Israel was said to be behind hits on Imad Moghnyeah, a chief Hezbollah operative, his son Jihad, and the bombing of Iranian arms convoys passing through Syria, among others. In the IDF doctrine, these punitive actions served as a deterrence by punishment and a way to postpone a large-scale conflagration.[48]

If deterrence by punishment was a critical aspect of cumulative deterrence, deterrence by denial was another one. To counter Hezbollah's threat of rockets and missiles, Israel has developed anti-rocket and anti-missile interceptors to protect the civilian population. Built by the Israeli aerospace industry, the Arrow III operates an exo-atmospheric interceptor to intercept long-range missiles. The Iron Dome, Israel's second operational missile defense system was developed to counter heavy mortars and short-range rockets. Tested in combat situations, notably in Protective Edge, the system scored impressive results. The system proved to be 80–90 percent effective in intercepting the incoming short-range rockets from over 300 kilometers (180 miles) away and shooting down incoming surface-to-surface and air-to-surface-missiles in a drill simulating rocket bombardments of Israeli cities.[49]

In yet another tactic of deterrence by denial, the IDF has unveiled tailored responses to the situation on the border of Lebanon and the Gaza Strip. In order to deny Hezbollah an opportunity for cross-border raids and flat-trajectory fire at communities near the border, the military has been in the process of upgrading the existing walls and fences and constructing precipices. In Gaza, where Hamas used an extensive network of tunnels, the IDF, in cooperation with the United States, developed tunnel-sensing technology to detect underground passages. The United States invested $120 million in the joint project. A first of its kind, the system provides data deciphered by advanced algorithms, which allows the military to locate tunnel-digging operations. The system uses hundreds of sensors detecting the noise of underground construction. It also boasts sophisticated computer software that identifies real signals from false ones.[50]

[48] Amos Harel, "Israel Outs a Senior Hezbollah Target," *Haaretz*, October 27, 2017.

[49] MAX, Strategic Analysis: Israel's Anti-Missile Systems: The Best Defense is Once Again a Good Defense, https://www.max-security.com/israels-anti-missile-systems-the-best-defense-is-a-good-defense/; Raoul Heinrichs, How Israel's Iron Dome Anti-Missile System Works, *Military and Defense*, July 30, 2014, www.businessinsider.com/how-israels-iron-dome-anti-missile-system-works-2014-7; Avi Lewis, Israel's new anti-ballistic missile system 'phenomenal' in testing, *The Times of Israel*, April 1, 2015, www.timesofisrael.com/israel-successfully-tests-new-anti-ballistic-missile-system/.

[50] Amos Harel, Israeli Army Readies for Flare-ups Ahead of Work on Lebanon Border Wall, *Haaretz*, June 25, 2017; Doron Almog, *Parameters*; RT, the US to invest $120 million into

Implemented by Talpiot IDF training program for soldiers who demonstrated aptitude in sciences, the system was successfully tested near the Kerem Shalom Crossing. On October 29, 2017, the IDF imitated the system to blow up a tunnel, which the PIJ dug from Khan Younis into the vicinity of Kibbutz Kissufim. Khalid Al-Batsh, a senior official in the PJI, explained that the tunnel was constructed to kidnap Israeli soldiers and vowed to avenge the "martyrdom" of several senior commanders who died in the collapse.[51]

Finally, as part of the cumulative deterrence strategy, the IDF has developed a procedure of "deterrence signaling"; that is, spelling out the high cost of potential military adventurism. Major General Gadi Eisenkot, then in charge of the Northern Command, Colonel Gabi Siboni, and Major General Giora Eiland, a former National Security Council chief, articulated a credible communication known as the "Dahya doctrine." A reference to the heavy bombing of the densely populated Shiite quarter of Dahya in Beirut during the 2006 war, it promised a devastating response to a Hezbollah action. Deterrence theorists have argued that such signaling may result in countersigning by the enemy in a cycle known as "deterrence bargaining." They have warned that "deterrence bargaining," notably when signaling is backed up by limited actions to make the signals credible, may lead to an unwanted conflict. But having determined that Hezbollah underestimated the threat of retaliations in 2006, the IDF went all way out on signaling.

In the most recent round, the Israeli Air Force chief Amir Eshel promised that in a future conflict there would be "potential power unimaginable in its scope, much different to what we have seen in the past and far greater than people estimate. What took us 36 days, now it would take 48–60 hours." He added that "if war breaks out in the north, we have to open with all our strength from the start, I have a recommendation to the people of Lebanon: Hezbollah has chosen to stay and infiltrate some

Israeli tunnel detection system – report, February 4, 2016, https://www.rt.com/news/331331-us-israel-tunnel-detection/.

Daniel Bernstein, Israel deploying revolutionary anti-tunnel system on Gaza border, *The Times of Israel*, April 16, 2015, www.timesofisrael.com/tunnel-detection-system-to-be-deployed-on-gaza-border/.

[51] Asher Zeiger, IDF to introduce new sensor system to locate tunnels in Gaza, *The Times of Israel*, July 2, 2012, www.timesofisrael.com/new-technology-to-help-idf-locate-tunnels-in-gaza/; Dov Lieber, "Exploded Tunnel Was Dug for Soldiers Abduction," *Times of Israel*, October 2017.

of your homes. If you leave your homes as soon as the conflict erupts, you will not be harmed."[52]

Not to be outdone, Sheikh Hassan Nasrallah threatened that thousands of volunteers would help Hezbollah fight the next war against Israel. "If Israel starts another war against Lebanon or Syria, hundreds of thousands of Muslims from other Islamic countries will come to the help of Lebanon and Syria to fight this regime." He noted the "Israeli enemy must know that if it launches a war against Syria or Lebanon, it will not be clear that whether this war will remain a Lebanese–Israeli or Syrian–Israeli conflict. Of course, this does not mean that other countries will directly get involved. But such a war will open the way for the participation of thousands, and even hundreds of thousands of volunteer fighters from all over the Arab world and Islam from Iraq, Yemen, Iran, Afghanistan, and Pakistan."[53]

Hezbollah's threat to unleash a massive attack on Israel was not an idle boast as it was closely related to the end-stage positioning in the Syrian civil war.

The End-Game in Syria: Israel's Gains and Losses

With most of Syria divided into de-escalation influence zones, the Golan Heights have witnessed an intense competition between Iranian-sponsored forces and groups aligned with the United States and Israel. The IRGC-QF made little secret of its desire to create a new anti-Israeli front in the strategic Heights, a gateway to Israel's vulnerable Sea of Galilee region. In the words of the Guards' strategist Hassan Abbasi, this was part of the so-called Sword of Damocles. He explained that "in the myths of ancient Greece, the sword of Damocles is a sword that is laid at the enemy's neck, which, if it makes any mistake, will cut off the head of its dear one. In the Doctrine of the Sword of Damocles, the situation in Syria, Islamic resistance in Palestine and Lebanon is in a way that the Sword of Damocles is placed around the neck of the Zionist regime from three angles."[54]

[52] AFP "Air Force Chief Boasts of unimaginable power in Future Lebanon War," *The Times of Israel*, June 21, 2017; Jean-Loup Samaan, "From War to Deterrence? Israel-Hezbollah Conflict since 2006," *Current Politics and Economics of the Middle East*, July 1, 2015.

[53] Safhe Aval, Nasrallah: in the Next War, Israel Should be Waiting for Hundreds of Thousands of Muslims from all Over the World, ifpnews.com/fa/exclusive/باشد-منتظر-بعدی-جنگ-اسرائیل/.

[54] Hassan Abbasi, Syria: Islamic World's Strategic Depth, in Mohammad Khalilpour, A Different Story of Logic of Being in Syria (Tehran: Masir-e Aftab, 2017).

To implement the Sword of Damocles, the Southern Shield Brigade is made up of Palestinians, Druze, Circassians and local Syrians, who were posted in the Hermon villages. The Brigade was commanded by Majd a-Din Khalik Khaymoud, who was killed in an ambush near the village of Khan Arnabah. The Israeli intelligence revealed that Munior Ali Naim Shaiti, known as Hajj Hasham, the head of Hezbollah operations in south Syria, was the head of the new front. Hajj Hasham was said to oversee a sizable group of Hezbollah and foreign fighters whose task it was to create cells capable of crossing into the Israeli held territory.[55]

While officially neutral in the conflict, Israel has secretly extended help to the local anti-Assad fighters. For several years, the IDF has overseen humanitarian aid dispensed from several centers near Quneitra. According to sources, the Israelis have also supported some local anti-Assad militias such as Fursan al Joulan which is said to have some 400 fighters. Moatasem al Golani, a spokesman for the group, admitted receiving help from Israel, which has been estimated at some $5,000 a month. Al Joulan fighters serve as a buffer zone between the Iran leaning units and provide intelligence on the latter's moves.[56]

Going beyond the Golan Heights, the IDF forces carried numerous actions, principally bombing of Hezbollah targets and convoys of weapon destined for the militia. Israeli politicians have repeatedly declared that they would disrupt the land bridge arms traffic to Lebanon. Despite sharp protests by the Assad government and his Iranian patrons, the Russian guarantors of the Astana geopolitical division of Syria, have not reacted. The ambiguity in Moscow's position which, as noted in Chap. 3, stemmed from its multifaceted agenda in Syria and beyond, has shown the limitation of the IRGC-QF victory. The Trump administration plans to declare the Guards as a terror group has added another constraint to its anti-Israel grand plan in Syria. Paradoxically but unsparingly, it helped the Rouhani normalizers notch up another victory for the parastatals and their patron the Supreme Leader. As the concluding chapter discusses, the competition between the normalizers and the traditionalists would continue to shape the country's foreign policy in the Middle East and beyond.

[55] Anna Ahroheim, "Israel Unmask New Hezbollah Leader on Syria Border," *Jerusalem Post*, October 25, 2017, www.jpost.com/Israel-News/Israel-unveils-new-Hezbollah-leader-on-Syria-border-508368.

[56] Rory Jones, Noam Raydan, Suha Maayeh, "Israel Gives Secret Aid to Syrian Rebels," *Wall Street Journal*, June 18, 2017.

Conclusion

From its inception, the Islamist Republic had targeted Israel on multiple levels. Ayatollah Khomeini declared Shiite Islam to be the new liberator of Jerusalem and the Holy Land from the so-called "Zionist entity," or Little Satan. Symbolically, Khomeini's dictum is celebrated annually during the Quds Liberation Day. The regime had consistently denied Israel's right to exist and harshly attacked Egypt and Jordan for making peace with Israel.

Helped by the extreme theology of the Haqqani School, Mahmoud Ahmadinejad ratcheted up the hostility to Israel. Having previously declared Iran to be part of the "the nuclear club," the president implied that nuclear weapons could be used to achieve this goal.

As part of the normalization plan, Rouhani tried to balance the theological dictates with a dose of realpolitik. The president and his circles admonished the hardliners for threatening Israel with extinction and for engaging in Holocaust denial. These were actions that in their view besmirched the international reputation of Iran.

For the Principalists, however, the rhetorical vilification of Israel was not only theologically mandated, but a platform for scoring points against the moderates. Hardline media did not change their tone when referring to the "Zionist enemy." They also accused the normalizers of being "soft" on Israel. Just to make their defiance of the Rouhani government clear, the Revolutionary Guards painted the slogan "death to Israel" on missiles paraded during a drill. Rouhani wrote a letter to Ayatollah Khamenei demanding the removal of the slogans, but the Supreme Leader who used the same line in his speeches, was not persuaded.

On the policy level, the normalizers wanted to keep Hezbollah, Hamas, and the Palestinian Islamic Jihad on a tight leash to avoid a conflagration with Israel, especially during the sensitive JCPOA negotiations. Albeit for different reasons, the hardliners shared this goal. With Hezbollah fully engaged in Syria, the Revolutionary Guards could hardly afford to open a new front against Israel. This restrain was more remarkable given that Israel bombed Iran's supplies of weapons to Hezbollah and Hezbollah's positions in Syria. But inadvertent developments can upset the delicate balance and trigger a new conflict, which neither the moderates nor even the hardliners in Iran desire.

References

AFP "Air Force Chief Boasts of unimaginable power in Future Lebanon War," *The Times of Israel*, June 21, 2017.

Akbar Hashemi Rafsanjani, Transcript of Rafsanjani Speech at Friday Prayer, Voice of the Islamic Republic of Iran Radio, December 14, 2001, http://www.globalsecurity.org/wmd/library/news/iran/2001/011214-text.html.

Ali M Ansari, Iran Under Ahmadinejad: The Politics of Confrontation (London: Routledge, 2007.

Amir Taheri, The Persian Night: Iran Under the Khomeinist Revolution (New York: Encounter Books, 2010.

Amir Taheri, "Hezbollah and Iran? Who Pays the Price," *Asharq al Awsat*, June 10, 2011.

Amir Taheri, Sympathy from the Devil. *New York Post*, September 23, 2007.

Ammariyon, The Victory of Lebanon's Resistance in 33-day war was Obtained by Iran's Support, *Ammariyon.ir*, http://www.ammariyon.ir/fa/news/18178/ پیروزی-مقاومت-لبنان-در-جنگ-33-روزه-با-حمایت-ایران-حاصل-شد.

Amos Harel and Gili Cohen, Hezbollah From terror group to army, *Haaretz*, June 12, 2016, www.haaretz.com/st/c/prod/eng/2016/07/lebanon2/.

Amos Harel, "Israel Launches Preemptive Strike in Psychological War with Hezbollah," *Haaretz*, June 24, 2017.

Amos Harel, "Israel Outs a Senior Hezbollah Target," *Haaretz*, October 27, 2017.

Amos Harel, Israeli Army Readies for Flare-ups Ahead of Work on Lebanon Border Wall, *Haaretz*, June 25, 2017.

Anna Ahroheim, "Israel Unmask New Hezbollah Leader on Syria Border," *Jerusalem Post*, October 25, 2017, www.jpost.com/Israel-News/Israel-unveils-new-Hezbollah-leader-on-Syria-border-508368.

Aparat, Interview with Hassan Rouhani in Qods Day, *Aparat*, www.aparat.com/v/nCNby/مصاحبه_حسن_روحانی_در_روز_قدس_سال_96.

Asher Zeiger, IDF to introduce new sensor system to locate tunnels in Gaza, *The Times of Israel*, July 2, 2012, www.timesofisrael.com/new-technology-to-help-idf-locate-tunnels-in-gaza/.

Avi Lewis, Israel's new anti-ballistic missile system 'phenomenal' in testing, *The Times of Israel*, April 1, 2015, www.timesofisrael.com/israel-successfully-tests-new-anti-ballistic-missile-system/.

AW, "Iranian Advisers Influence Course of Lebanon/Israel Conflict". *Aviation Week*, August 14, 2006. Archived from the original on 29 April 2011.

BI, Satellite imagery reveals Hezbollah's play for permanent bases in Syria, *Business Insider*, April 6, 2016, www.businessinsider.com/satellite-imagery-reveals-hezbollahs-play-for-permanent-bases-in-syria-2016-4.

Dan De Luce, Syrian War Takes Rising Toll on Hezbollah, *Foreign Policy*, July 9, 2015, foreignpolicy.com/2015/07/09/syrian-war-takes-rising-toll-on-hezbollah/.

Daniel Bernstein, Israel deploying revolutionary anti-tunnel system on Gaza border, *The Times of Israel*, April 16, 2015, www.timesofisrael.com/tunnel-detection-system-to-be-deployed-on-gaza-border/.

Daniel Sobelman, "Learning to Deter: Deference Failure and Success in the Israeli-Hezbollah Conflict", 2006–2016.

Doron Almog, "Cumulative Deterrence and the War on Terrorism," *Parameters*, Winter, 2004–2005.

Doron Almog, Parameters: RT, the US to invest $120 million into Israeli tunnel detection system – report, February 4, 2016, https://www.rt.com/news/331331-us-israel-tunnel-detection/.

Dov Lieber, "Exploded Tunnel Was Dug for Soldiers Abduction," *Times of Israel*, October 2017.

DW, Iran Denied Sending Fajr-5 Missile to Hamas, *DW*, November 18, 2012, www.dw.com/fa-ir/ایران-ارسال-موشک-فجر۵--برای-حماس-را-تکذیب-کرد/a-16387880.

Entekhab, Iran Declared: Hezbullah Drone was Iranian, *Entekhab.ir*, www.entekhab.ir/fa/news/80118/ایران-تایید-کرد-هواپیمای-بدون-سرنشین-حزب-الله-ایرانی-بوده-است.

Ethan Bronner, Just How Far Did They Go, Those Words Against Israel? *The New York Times*, June 11, 2006.

Ewen MacAskill and Chris McGreal, Israel Should Be Wiped Off Map, Says Iran's President. *The Guardian*, October 27, 2005.

Fallah, Israel Nature from Imam Khomeini Perspective, August 17, 2012, www.al-falah.ir/portal/?pageID=910.

Fars-News-Agency, Ahmadinejad Remarks on Zionist Regime Provoked Wrath of Zionism Supporters. *Fars News Agency*, October 27, 2005, http://www.farsnews.com/newstext.php?nn=8408040372.

FNA, Commander-in-Chief of the Guards: Iran did not send weapons to Gaza, but Technology Transfer, *Fars News Agency*, www.khabaronline.ir/detail/259262/Politics/military.

Gilli Cohen, "Israeli Army Intelligence Chief: Hezbollah Is Setting Up a Weapons Industry in Lebanon With Iranian Know-how," *Haaretz*, June 23, 2017.

Hanin Ghaddar, The Marriage and Divorce of Hamas and Hezbollah, *Wilson Center*, https://www.wilsoncenter.org/article/the-marriage-and-divorce-hamas-and-hezbollah.

Hassan Abbasi, Syria: Islamic World's Strategic Depth, in Mohammad Khalilpour, A Different Story of Logic of Being in Syria (Tehran: Masir-e Aftab, 2017).

Hazem Balousha "Why Hamas resumed ties with Iran," *Al-Monitor*, June 29, 2016.

Iran Caricator, The opening ceremony of the 2nd International Cartoon Contest and the Holocaust Cartoon, May 14, 2016, art.irancartoon.ir/content/مراسم-افتتاح2-مین-مسابقه-بین-المللی-کارتون-و-کاریکاتور-هولوکاست-برگزار-شد/.

Iran Caricature, The Opening Ceremony of the 2nd Holocaust International Cartoon Contest, *Iran Caricature*, May 14, 2016, art.irancartoon.ir/content/مراسم-افتتاح2-مین-مسابقه-بین-المللی-کارتون-و-کاریکاتور-هولوکاست-برگزار-شد/.

IRARTESH, Zelzal Rocket the most accurate Iranian rocket, February 19, 2013, irartesh.ir/post/581/ایرانی_راکت_ترین_دقیق_زلزال.html.

IRDC, Darbareye Roze Jahaniye Qods [About International Day of Qods]. *Islamic Revolution Documents Center*, July 23, 2014, http://www.irdc.ir/fa/content/47167/print.aspx.

IRNA, Ahmadinejad: Those Who Recognize Israel Will be Punished by the Wrath of the Islamic Nations. *Islamic Republic News Agency*, October 26, 2005, http://www.president.ir/fa/2288.

IRNA, Bayanat dar Didar-e Zaerin-e Haram-e-Motahar Razavi [Speech in Mash'had for Imam Reza Pilgrims]. *Islamic Republic News Agency*, March 22, http://news.iran-emrooz.de/more.php?id=11566_0_7_0_M.

Israel's war With Iran by Ze'ev Schiff, *Foreign Affairs*, November–December 2006.

Ivan Arreguín-Toft, "Unconventional Deterrence: How the Weak Deter the Strong," in Paul, Morgan, and Wirtz, Complex Deterrence.

Ivan Arreguín-Toft, How the Weak Win Wars: A Theory of Asymmetric Conflict (Cambridge: Cambridge University Press, 2005).

IW, The Holocaust Cartoons and Zarif's Lies, *Iran Wire*, https://iranwire.com/en/features/1781.

Jame'e Khabar, Iran Called "Death to Israel", April 18, 2017, https://www.jamehnews.com/8419/روزنامه-انگلیسی-ایران-در-روز-ارتش-شعار-مرگ-بر-اسرائیل-سر-داد-1.

JDW, "Iran answers Hizbullah call for SAM systems." *Jane's Defense Weekly*, 2006-08-07.

Jean-Loup Samaan, "From War to Deterrence? Israel-Hezbollah Conflict since 2006," *Current Politics and Economics of the Middle East*, July 1, 2015.

Jodi Vittori, Terrorist Financing and Resourcing (Palgrave Macmillan, 2011).

John R, Davis, Jr, "Defeating Future Hybrid Threats: The Greatest Challenge to the Army Profession of 2020 and Beyond," *Military Review*, September–October, 2013.

Judah A. Gross, IDF chief: Hezbollah commander killed by his own men last year, *Times of Israel*, March 21, 2017, www.timesofisrael.com/idf-chief-hezbollah-commander-killed-by-his-own-men-last-year/.

Khamenei, Bayanat dar Khotbe'haye Namaz Jom'e Tehran (Speech at Tehran Friday Prayers). *Khamenei.ir*, October 21, 2005.

Khamenei.ir, Khamenei's tweet on September 9, 2015, https://twitter.com/khamenei_ir/status/641520641724780544?ref_src=twsrc%5Etfw&ref_url=https:%2F%2Fwww.timesofisrael.com%2Fkhamenei-israel-wont-survive-next-25-years%2F.

Khamenei: Zionist Regime is Doomed to Fail, *Khamenei.ir*, July 9, 2015, farsi.khamenei.ir/photo-album?id=24577.

Laboratory of Asymmetry: The 2006 Lebanon War and the Evolution of Iranian Ground Tactics Lindemann, Marc, Military Review May–June 2010.

Learning to Deter: Deterrence Failure and Success in the Israeli-Hezbollah Conflict. *International Security*, 41, issue 3, 2017.

ManotoTV, IRGC Commander Called for Readiness for War with Israel, September 22, 2012, https://www.youtube.com/watch?v=GFtYnk6a3ys.

Matthew Levitt, Hamas: Politics, Charity, and Terrorism in the Service of Jihad (Yale University Press, 2006).

MAX, Strategic Analysis: Israel's Anti-Missile Systems: The Best Defense is Once Again a Good Defense, https://www.max-security.com/israels-anti-missile-systems-the-best-defense-is-a-good-defense/.

MNA, Zarif; We got comfort from Netanyahu, *Mehr News Agency*, May 6, 2014, www.mehrnews.com/news/2286013/رمز-دلواپسی-قدوسی-کریمی-ایم-گرفته-نتانیاهو-از-را-آسایش-ظریف.

Nadav Pollak, The Transformation of Hezbollah by its Involvement in Syria (Washington, DC: Washington Institute for Near East Policy, 2015).

Nazila Fathi, Wipe Israel 'Off the Map' Iranian Says. *The New York Times*, October 27, 2005.

Qor'an, Surat Al-Isra (The Night Jurney), 17:104, https://quran.com/17.

Qur'anic Promise of Annihilation of Israel, https://www.tasnimnews.com/fa/news/1396/04/01/1443455/-زمان-درباره-تفسیری-نظریات-آخرالزمان-در-صهیونیسم-نابودی-قرآنی-وعده اسرائیل-عذاب-بنی-تحقق.

Rahkar News, Rouhani Called for Removing "Death to Israel" Writing on the Missiles, March 26, 2016, www.rahkarnews.ir/fa/doc/news/20794/روحانی-خواستار-حذف-شعار-هک-شده-مرگ-اسرائیل-روی-موشک-های-بالستیک.

Raoul Heinrichs, How Israel's Iron Dome Anti-Missile System Works, *Military and Defense*, July 30, 2014, www.businessinsider.com/how-israels-iron-dome-anti-missile-system-works-2014-7.

Reuters. 2008. Ahmadinejad Says Israel Will 'Disappear'. *Reuters*, June 3, http://www.reuters.com/article/2008/06/03/us-iran-israel-usa-idUSL0261250620080603.

RFI, The Second Round of Rejection of the Killing of the Jews, January 27, 2009, www1.rfi.fr/actufa/articles/109/article_5152.asp.

Richard Davis, Hamas, Popular Support and War in the Middle East: Insurgency in the Holy Land (New York: Routledge, 2016).

Robin Wright, Iran's Javad Zarif on the Fraying Nuclear Deal, U.S. Relations, and Holocaust Cartoons, *The New Yorker*, April 25, 2016, www.newyorker.com/news/news-desk/irans-javad-zarif-on-the-fraying-nuclear-deal-u-s-relations-and-holocaust-cartoons.

Roi Kais, Hezbollah has fleet of 200 Iranian-made UAVs, November 25, 2013, www.ynetnews.com/articles/0,7340,L-4457653,00.html.

Rory Jones, Noam Raydan, Suha Maayeh, "Israel Gives Secret Aid to Syrian Rebels," *Wall Street Journal*, June 18, 2017.

Rouhollah.ir, Sahifeh Imam Khomeini, Vol. 15.

Ruhollah Khomeini, Sahifeh Noor, Vol. 14.

Ruhollah Khomeini, Sahifeh Noor, Vol. 4: Khomeini.ir, Imam Khomeini and Camp David Plot, September 19, 2012, www.imam-khomeini.ir/fa/NewsPrint.aspx?ID=3062.

Safhe Aval, Nasrallah: in the Next War, Israel Should be Waiting for Hundreds of Thousands of Muslims from all Over the World, ifpnews.com/fa/exclusive/باشد-منتظر-بعدی-جنگ-اسرائیل/.

Shafaghna, War With Israel Would Soon Break out, Shia News Association, September 22, 2012, www.shafaqna.com/persian/component/k2/item/22997-فرمانده-سپاه-دشمني-رژيم-صهيونيستي-با-ايران-به-درگيري-فيزيكي-ختم-مي‌شود.

Sharq Newspaper, "Interview with Ali Akbar Mohtashami." *Shargh Newspaper*, 2006-08-03, ww25.sharghnewspaper.com/850512/html/polit.htm#s453504.

Shiite News, "Islamic Resistance," *Shiite News*, September 14, 2015.

Shimon Shapira "Hezbollah's Veneration of Iranian Leader Ali Khamenei." *Jerusalem Center Blog*, June 20, 2011.

Simon Mabon, Saudi Arabia and Iran: Power and Rivalry in the Middle East (I.B. Tauris, 2016.

Speaking for the 50 thousand Nationwide Basij Commanders, *Khamenei.ir*, farsi.khamenei.ir/speech-content?id=24552.

Stratfor, "Hezbollah's Shot at Permanency in Syria," *Stratfor*, April 6, 2016, https://www.stratfor.com/analysis/hezbollahs-shot-permanency-syria.

The organizers increased the winning awards from $ 20,000 to $50,000 Holocaust Denial: Iran Holocaust Cartoon Exhibition, https://www.ushmm.org/fa/iran-cartoon-fa.

Times of Israel Staff, "Unveiling clock showing 8,411 days left for Israel, Iranians rage against Jewish state," *Times of Israel*, June 24, 2017.

TNA, Reviewing how the 'Israeli Mercado Tank was Destroyed' at the University of Revolutionary Guards, https://www.tasnimnews.com/fa/news/1396/05/01/1467751/پاسداران-سپاه-افسری-دانشگاه-در-اسرائیل-مرکاوای-تانک-انهدام-نحوه-بررسی.

TNA, The End of Four-Percent Government, *Tasnim News Agency*, May 6, 2017, https://www.tasnimnews.com/fa/news/1396/02/16/1399228/پایان-دولت4--درصدی-ها.

Tnews, Rouhani Speech Gives excuse to the West, tnews.ir/news/fab485219452.html.

TNYS, "War Dead Flown to Iran". *The New York Sun*, July 24, 2006. Archived from the original on January 4, 2011.

Uzi Mahnaimi, Security shambles as Hezbollah drone spies on Israeli army, *The Times*, October 14, 2012, https://www.thetimes.co.uk/article/security-shambles-as-hezbollah-drone-spies-on-israeli-army-n6kxp77dnz7.

Yaakov Katz, (August 4, 2006). "Iran: We supplied Zelzal-2 to Hizbullah". *Jerusalem Post*. Archived from the original on July 6, 2013. Retrieved 2006-08-13.

Ynet, Report: Iranian soldiers join Hizbullah in fighting, *Ynet News*, July 24, 2006, www.ynetnews.com/Ext/Comp/ArticleLayout/CdaArticlePrintPreview/1,2506,L-3280446,00.html.

Yonah Jeromy Bob, "Massive Iranian Funding for Anti-Israel Terror Group Revealed," *Jerusalem Post*, June 22, 2017.

Zarif, personal Twitter Account, September 5, 2013, https://twitter.com/jzarif/status/375617854214660097?lang=en.

CHAPTER 10

Conclusions

From its inception, the Islamic Republic of Iran struggled with two hard-to-reconcile approaches to foreign policy. Ayatollah Ruhollah Khomeini, the "Iranian Trotsky" who wanted to export the revolution to Muslim countries occupied the idealist part of the spectrum. His chief lieutenant, Akbar Hashemi Rafsanjani was much more of a realist. He believed that the new republic should concentrate on economic viability for which a more traditional foreign policy was required. Making the idealist-realist schism worse was the negotiated political order whereby the Revolutionary Guards and other parastatals operated virtually outside the purview of the state. The resulting opaqueness and confusion of the foreign policy-making process have vexed policymakers, academics, and lay observers alike. Still, the detailed analysis of the major foreign policy issues has uncovered specific patterns that should help to decipher this complex system.

Depending on the area, it is either the state or the parastatals that dominate the policy-making process. The highly secretive Rahbar Office operates behind the scene to monitor the discourse and side the state normalizers or the hardline parastatals. Complicating matters, under President Mahmoud Ahmadinejad, a Principalist, the state, normally a preserve of the realists, has adopted the idealistic positions of Khomeini's revolutionary export. But the political shift of Ahmadinejad did not produce a unanimous decision-making process. Anxious about the legitimacy crisis provided by the nuclear sanctions, the Supreme Leader Ayatollah Ali Khamenei sided with the normalizers, a move that paved Hassan Rouhani's way to the office.

F. Rezaei, *Iran's Foreign Policy After the Nuclear Agreement*, Middle East Today, https://doi.org/10.1007/978-3-319-76789-5_10

Having to operate within the confines of the negotiated political order forced regime players to compete over every foreign policy issue. To a large extent, this competition has shaped the approach described in the period covered by this work. By and large, the normalizers in the government have had a dominant input into the policy toward the United States, Russia, the EU, and Turkey. Having embraced a substantially realistic approach to international relations, they were open to the so-called Iran outreach. A long-standing notion in American liberal circles, it was premised on the understanding that it was possible "to do business" with the moderates in Tehran and thus help them to undermine the hardliners.

The Obama administration encouraged the normalizers to engage in the nuclear negotiations and promised to reward them with sanction relief. The EU, as noted, had a history of dealing with the moderates, which preceded the election of Rouhani. Like the Obama administration, the EU was eager to prove that working with the moderates would move the regime ever so incrementally toward normalization and engagement in the international order. The EU thus saw the JCPOA as a double win. It opened Iran for European business and promised to unleash the dynamics for a peaceful regime change.

Russia and Turkey were mainly under the purview of the statist-moderate domain, albeit for a different reason. There was little disagreement between the normalizers and hardliners over Russia, a veteran Iran partner in the nuclear project, notwithstanding Ayatollah Khomeini's reservation of conducting relations with a "godless" enemy. Indeed, both the normalizers and the hardliners were keen on pursing the pivot to the East with Russia as its linchpin. Although the hardliners clashed with Turkey on an array of issues, including Syria, the common interests in fighting the Kurds paved the way for a joint approach with the normalizers who appreciated Turkey's help with the sanctions and the considerable potential of future economic relations.

Relations with Iraq, Syria, and Israel are located on the other end of the spectrum and have been dominated by the traditionalists, a coalition of Principalists including the Revolutionary Guards, and the Haqqani School, which enjoyed the considerable support of the Rahbar Office. As Khomeini-style idealists, the traditionalists believe that it was the duty of the Islamic Republic to spread the revolution and challenge the state of Israel. To carry out these twin missions, the IRGC-QF created proxies modeled after the Lebanese Hezbollah.

So successful was the Lebanese model, a mixture of political and military control of Lebanon, that the Guards rushed to implement it in the post-2003 Iraq. Bypassing the government, the IRGC tried to achieve the "Lebanon effect" through infiltrating the Shiite dominated political process and forging a strong Shiite paramilitary force. The Guards even subverted the traditional tools of statecraft since many top Iranian diplomats in Iraq were former Guards commanders and consular offices around the country were similarly staffed. Sidelined by the Guards, the Rouhani government had to accept Qassem Suleimani as its de facto representative in Iraq. It was only after the ISIS debacle—which the normalizers blamed on Suleimani—that normalizers asserted partial and tenuous control in Iraq.

It was natural for the IRGC, which had long-standing relations with the Syrian regime, to come to Assad's rescue during the Arab Spring. While the hardliners considered the "defense of the shrine" akin to the Iran–Iraq war, the normalizers criticized the considerable cost in blood and treasure a waste of Iran's limited resources. What is more, in a historic turn, the moderates decried the revolutionary export as ruinous adventurism in foreign policy. Like in Iraq, the normalizers regained a measure of control over the Syrian policy because of the failure of the IRGC-QF to prop up Assad without the help of Russia. Ironically, the negotiations in Astana, which featured more traditional diplomacy, gave the Rouhani government a somewhat bigger role in Syrian policy. But the Astana agreement did not diminish the power of the Revolutionary Guards on the ground. To the contrary, with the last ISIS redoubts on the Syrian–Iraqi border about to fall in the hands of Syrian government forces and the Shiite militias, the Revolutionary Guards recovered their footing and leverage against the normalizers.

Israel, however, the fulcrum of Khomeini's Islamist mission, was a harder terrain to penetrate. As the book demonstrates, the normalizers were acutely aware that extravagant anti-Israel rhetoric had a detrimental effect on Iran's international image. But they could do little to tone down the discourse, not least because the Supreme Leader was behind the theory of Israel's alleged demise. The Rouhani government was equally powerless to influence the Revolutionary Guards plans to create a new front against Israel. More critically, the Israeli Defense Force (IDF) bombing raids in Syria have forced the president and his foreign minister to condemn Israel, creating the impression that the normalizers are as bitter enemies of the Jewish states as the hardliners.

Located between the two poles of the spectrum was Saudi Arabia where both the normalizers and the hardliners competed in shaping its foreign policy. While Ayatollah Khomeini considered Saudi Arabia an archenemy of his brand of revolutionary Islam, pragmatists like Ayatollah Rafsanjani, Mohammed Khatami, and Rouhani urged to mend relations with their powerful Gulf neighbor. But the IRGC-QF decision to assist the Houthi rebels, a choice supported by Ayatollah Khamenei as a part of his plan to create the A2/AD maritime defense system, was hardly helpful in this context. King Salman's tougher line toward Iran all but squashed the idea.

Also, broadening extant scholarship, the analysis of foreign policymaking in a negotiated political order helps with the applied concerns of sanctions statecraft. Despite being an object of a variety of sanctions since 1979, the impact of sanctions on Iran has never been well understood. Indeed, the field was so contentious that it contributed to the emergence of two schools of thought, the so-called sanction optimists, and sanction pessimists. The JCPOA was an intellectual victory for the former who asserted that sanction statecraft would be enough to roll back Iran's nuclear project. As noted, the Supreme Leader sided with the moderates out of fear that the deteriorating economy would doom the legitimacy of the regime. Regarding the negotiated political order, the normalizers gained dominance when the Principalist position became too costly for the entire regime. In other words, rather than weaken the moderates, an assumption made by sanction pessimists, the economic pain added to their strength.

While less comprehensive than the nuclear sanctions, the measures that the Trump administration has contemplated against IRGC-QF should create the same dynamics. In fact, the Rouhani government has been empowered enough to start a campaign against companies owned by the Guards, accusing them of corruption and maleficence. Because of the extent of the new sanctions, normalizers could make a compelling argument that the IRGC extensive ownership of assets is creating a drag on the economy. Interestingly, Ayatollah Khamenei, still anxious about the fragile economic recovery, has not addressed the issue. Should the economic damage from the new sanctions be extensive, the Supreme Leader may be forced to side with the normalizers' push to curtail the hardline revolutionary exporters.

Index[1]

[1] Note: Page numbers followed by 'n' refer to notes.

F. Rezaei, *Iran's Foreign Policy After the Nuclear Agreement*, Middle East Today, https://doi.org/10.1007/978-3-319-76789-5

F

G

H

I

T

U

The manufacturer's authorised representative in the EU is Springer Nature Customer Service Centre GmbH, Europaplatz 3, 69115 Heidelberg, Germany. If you have any concerns regarding our products, please contact ProductSafety@springernature.com

Printed and bound by CPI Group (UK) Ltd, Croydon, CR0 4YY
05/10/2026
02228140-0004